Japanese Agriculture

Japanese Agriculture
Patterns of Rural Development

Richard H. Moore

Westview Press
BOULDER, SAN FRANCISCO, & LONDON

H-Y LIB
(W)

HD
920
.N33
M65
1990x

Photographs courtesy of Richard H. Moore.

Westview Special Studies on East Asia

This Westview softcover edition is printed on acid-free paper and bound in library-quality, coated covers that carry the highest rating of the National Association of State Textbook Administrators, in consultation with the Association of American Publishers and the Book Manufacturers' Institute.

Published in 1990 in the United States of America by Westview Press, Inc., 5500 Central Avenue, Boulder, Colorado 80301, and in the United Kingdom by Westview Press, Inc., 13 Brunswick Centre, London WC1N 1AF, England

Library of Congress Cataloging-in-Publication Data
Moore, Richard H.
   Japanese agriculture: patterns of rural development.
   (Westview special studies on East Asia)
   Bibliography: p.
   1. Land tenure—Japan—Nakada-chō (Miyagi-ken)
2. Agriculture and state—Japan—Nakada-chō (Miyagi-ken)
3. Rural development—Japan—Nakada-chō (Miyagi-ken)
4. Rice—Japan—Nakada-chō (Miyagi-ken)   I. Title.
II. Series.
HD920.N33M65   1990   338.1'0952'115   87-13553
ISBN 0-8133-7406-5

Printed and bound in the United States of America

The paper used in this publication meets the requirements of the American National Standard for Permanence of Paper for Printed Library Materials Z39.48-1984.

10   9   8   7   6   5   4   3   2   1

Dedicated to my brother
Thomas Gordon Moore

Contents

viii

8   NŌKYŌ: THE AGRICULTURAL COOPERATIVE   137

9   INDUSTRIAL DEVELOPMENT AND THE RISE OF RURAL SUBCONTRACTING. 171

10   THE SCALE OF RICE FARMING AND RESISTANCE TO THE LAND IMPROVEMENT PROJECT 199

Tables, Figures, and Photos

TABLES

xiv

FIGURES

xvi

PHOTOS

Acknowledgments

I gratefully acknowledge the financial support of the Japan Foundation for my field research conducted between 1982 and 1983. I am also indebted to the Ohio State University for a seed grant to continue the research during the summer of 1987.

Many people have helped me in this study and space does not permit me to include them all. However, I do wish to thank a few persons and groups without whom this research would not have been possible.

As a graduate student at the University of Texas at Austin I was fortunate to have had Dr. John B. Cornell as my Ph.D. committee chairman. He urged me to examine Japanese land tenure and devoted many hours helping me to formulate my ideas.

While engaged in fieldwork, I was affiliated with the Office of Agricultural Economics at Tōhoku University Institute for Agricultural Research. The guidance and friendship of this research group, headed by Dr. Kazushige Kawai, was indispensable. Fellow graduate and postgraduate researchers Satoshi Arai and Hiroshi Owada instructed me on the fundamentals of Japanese agricultural economics. I am also indebted to Shigeru Sugiyama of the National Research Institute for Agricultural Economics and Rural Sociology, Masayoshi Namiki and Toshimune Hayashi of the Food Supply and Agricultural Policy Research Center, Takehiko Ohkawa of the Miyagi Prefecture Land Commission, Dr. Kunio Sasaki of Miyagi Agricultural College, and Dr. Tsuyoshi Honda of Miyagi Women's University.

My deepest appreciation goes to the people of Nakada Township and K Hamlet for their cooperation, patience, and international goodwill. Mayors Tetsu Haga and Gorō Miura and their staff from all sections of the township office were indispensable in providing information, introducing me to the local customs, and making my stay comfortable. K Hamlet will always be my home away from home.

Both the Miyagi Prefecture and Nakada Township Land Commission kindly showed me land tenure records and explained the intricacies of buying and selling farmland. The Land Improvement District staff members were extremely helpful in describing the dimensions of the Land Improvement Project.

I am very appreciative to Westview editor Kellie Masterson for realizing the need for the book and for convincing me to include Chapter 12 and to project editor Deborah Lynes for her attention to detail. Finally, I am very grateful to Scott Schnell, Nobuko Imai, Masako Kanno, and my wife, Nancy, for assistance in proofreading and the moral encouragement necessary to finish the project.

Richard H. Moore

1

Introduction

On September 14th, 1986 the United States Rice Millers Association (RMA) claimed in a petition filed under Section 301 of the 1974 Trade Act that the rice policies of Japan violate international trade laws and prevent fair and equitable access to the Japanese market. Critics argued that Japanese rice production costs are nearly six times greater than in the United States. Still unresolved in 1989, the issue has been pursued by the United States government from the vantage point of free trade while the Japanese government resisted on the grounds that rice self-sufficiency was important to national security.

This book takes a different approach from either of the above and instead stresses examination of the social impact of food and its production. Only after this has been understood can the larger issues of world trade be addressed, as I have attempted to do in Chapter 12. The focus needs to be redirected from the narrow logic of free trade rules or political convenience to the nutritive value of the Japanese diet, the ecology of food, and its cultural significance.

THE JAPANESE DIET

The overall Japanese diet has improved remarkably over the last thirty years. No doubt this has contributed to general health and longevity, of which Japan is a world leader. Despite the fact that rice consumption has been decreasing, average consumption in 1985 was about seventy-five kilograms (165 pounds). This was more than twice the wheat that was consumed

1

per capita. With rice as the central staple, the balance of protein, fat, and carbohydrate is close to the ideal set by the World Health Organization. Improvements in the Japanese diet are shown in Table 1.1. By slightly increasing the amount of protein and nearly quadrupling the dietary intake of fat per capita, during the period 1955 to 1985 Japan has been able to gradually achieve a nearly ideal PFC (protein/fat/carbohydrate) balance unsurpassed by other industrial countries. Table 1.2 shows Japan's PFC balance compared with other industrial nations.

THE ECOLOGY OF JAPANESE RICE AGRICULTURE

The ecology of Japanese rice production is a second consideration meriting attention. Japan, a country roughly 80 percent the size of California, is an archipelago where mountains make up three-fourths of its area. Rice has been Japan's central crop, with several rice belts predominating. In fact, rice was so well adapted to Japan that it quickly became the leading crop. Between 800 and 900 A.D. about one million hectares (see Appendix B for metric conversions) of rice were cultivated. By 1550, 1,500,000 hectares were under cultivation and by 1880 this amount rose to 2,600,000 hectares. Since 1900 the number of hectares has stayed about the same at three million hectares (Honda 1982:134). Only 38.6 percent of the paddy land is suitable for large-scale mechanization, which demands at least fifty hectare paddies with a slope of less than one percent (Tsuchiya 1976:85). Small-scale rice paddies molded to the contour of the natural environment have preserved and prevented erosion for over the two thousand years in which the Japanese have grown rice. Irrigation water, largely pumped from upstream to downstream, has usually followed the natural slope of the land. The average ten are (0.1 hectare) rice paddy is covered by over fifteen hundred tons of water, the major part of which becomes underground water.

Farmers traditionally maintained a combination of rice paddies and field crops so that their diet would be complemented by vegetables and other grains. A typical farmer held land in at least two environmental soil zones because the soil for rice is not usually well-suited for growing vegetables. Within the two environmental zones farmers commonly worked land in up to eight scattered locations, a result of fragmentation caused by

Table 1.1
Improvements in the Japanese Per Capita Dietary PFC Balance
(1955-1985)

Year	Protein (%)	Fat (%)	Carbohydrates (%)
1955	11.3	8.8	79.9
1960	12.2	11.4	76.4
1965	12.2	16.2	71.6
1970	12.4	20.0	67.6
1975	12.8	22.8	64.4
1980	12.9	25.5	61.6
1985	13.0	27.4	59.5
Ideal	12-13%	20-30%	57-68%

Source: Nōrintōkei Kyōkai 1986:47.

Table 1.2
The Dietary Per Capita PFC Balance in Industrialized Countries

Country	Protein (%)	Fat (%)	Carbohydrates (%)
Japan	13.0	27.4	59.5
United States	12.2	45.0	42.8
England	12.0	41.7	46.3
West Germany	11.4	47.8	40.8
France	13.3	46.9	39.8
Italy	13.0	40.1	46.9
Switzerland	12.5	45.2	42.3
Canada	12.7	43.6	43.7
Ideal	12-13%	20-30%	57-68%

Source: Nōrinsuisanshō Daijin Kanbō Kikakushitsu 1986:273.

Note: The data are based on 1982 statistics except for the data on Japan which is from 1985.

formation of new households from existing ones. While working their fragmented paddies, farmers became aware of micro-environmental differences between locations which often enabled farmers to time production and to take advantage of water allocation and drainage differences while fully utilizing household labor.

Rice cultivation offered more than soil conservation and food. Rice straw was used for making rope, tatami mats, paper; as food and bedding for cattle and horses; as thatching material for dwellings and other shelters; as fuel; as a mulch; as a source of organic matter in the soil; as a green fertilizer; as the container for certain food products such as nattō, and as the basic material from which all sorts of handicrafts were made. The rice chaff was used as fuel for cooking the rice and created that slightly browned bottom of cooked rice appreciated by Japanese. The paddies, before the introduction of herbicides and pesticides, hosted frogs, fish, and loaches (dojō), all of which were used for food. At harvest time locusts were gathered by the children and sweetened into a delicacy.

THE CULTURAL MEANING OF RICE

Rice cultivation is as old as the Japanese nation state and is recorded in the Nihon Shoki and Kojiki, the Shintō creation myth texts which date to before 720 A.D. Upholding the Nihon Shoki legend that Japan is the "Mizuho No Kuni" ("Land of Abundant Rice"), the emperor of Japan is the Shintō embodiment of Ninigi-no-mikoto, the God of the Ripened Rice Plant. Having entered the womb of the Sun Goddess (Amaterasu Ōmikami) during the Daijō-sai (Great Thanksgiving Festival)[1] that occurs upon enthronement, each emperor annually plants and harvests by hand a small plot of rice. This is offered to the Shintō gods at Ise Shrine on October 15 at the Kannamesai (Tasting by Dieties Festival) and November 23 at the Niinamesai (Festival of the New Tasting). These festivals, according to myth, were first performed by the divine imperial ancestress, the Sun Goddess, using newly harvested rice from the sacred fields in heaven. At dawn on New Year's Day farmers throughout Japan offer part of the autumn rice harvest to their hamlet Shintō shrine, in appreciation for last year's plentiful harvest.

Rice is a dominant symbol in Japanese society. The qualities of dominant symbols as described by Victor Turner (1967:28) include: (1) condensation; (2) unification of disparate meanings; (3) polarization of meaning. Condensation refers to the high emotional quality of the symbol. For example, in heated discussions about the issue of agricultural free trade, I have heard a few farmers end the discussion abruptly by quoting from the Kojiki that Japan is "the land of abundant rice," thus precluding the possibility of negotiation. Likewise, the sale of sekihan (red rice) which is served at festive occasions was curtailed during the Emperor Showa's prolonged critical illness.

Second, rice unifies disparate meanings which by themselves seem rather unconnected. For instance, fertility, maintaining the household line, land holdings, wealth, and old age are by themselves separate categories. Rice is a link between these and can symbolize any or all of them. In northern Miyagi Prefecture, where much of the data for this book was researched, women still meet twice a year to pray for fertility and family well-being. During their ritual they offer rice to be blessed by the mountain god. This rice is taken home to be cooked with the rice to be eaten at the family's evening meal. Likewise, rice and rice wine (sake) feature prominently in all Shintō weddings. In some rural weddings rice is prominently displayed to demonstrate succession of the household line and inheritance of land. It is also sent as zōtōmai (gift rice) to non-inheriting relatives who have moved to the city. In both rural and urban weddings sake is ritually served in three saucers and is sipped from each saucer three times because the number three carries auspicious meaning. Sake is also used to ritually purify oneself and the group and is customarily imbibed during shrine festivals as a collective representation of the neighborhood or hamlet group. Mochi (glutinous rice) was pounded in a wooden container in the traditional welcoming reception for visitors or during festive occasions. The eighty-eighth birthday in Japan is time for particular celebration with ritual drinking of sake. Also the numbers themselves, as Chinese characters written vertically closely together, appear to resemble the character for rice.

Rice relates to hamlet solidarity and an ideological sense of communality. Most hamlets in Japan, for instance, possess a Shintō shrine and hold an annual shrine festival in association with the shrine. The shrine serves as the focal point of hamlet egalitarian ideology and notion of hamlet territory. Most farmers possess a clear idea regarding their hamlet territory based on the location of rice paddy irrigation and drainage ditches. Ranging from 37 percent in Kagoshima Prefecture to 96 percent in Okayama Prefecture, the national average for hamlets having such territorial consciousness is 79 percent (Kawaguchi 1983). In many cases, hamlet cooperative labor is used to clean the ditches once or twice a year.

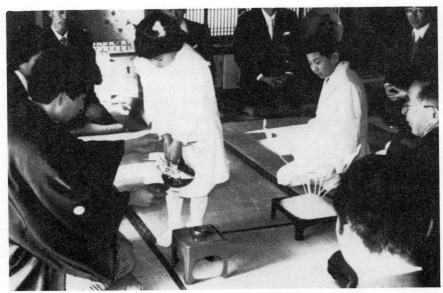

Photo 1.1
Rice as Symbol of Household Succession in Rural Wedding

Photo 1.2
Symbolic Use of <u>Sake</u> in Rural Wedding

Photo 1.3
Rice Offered to Hamlet Shrine at Daybreak on New Year's Day

Photo 1.4
Rice and _Sake_ Offered to Mountain Gods in K Hamlet

Last, rice serves to make clear the polarization of meaning in Japanese society. In the continuum of hare (purity) and kegare (impurity), rice is more closely associated with life and Shinto. Cosmologically this is opposite from the pollution of death that is closely associated with the Buddhist religion.

Yet rice can also be associated with Buddhism in that it can be ritually served to ancestors. For example, during the Feast of the Dead (Ōbon), which occurs in mid-August, dango (rice flour balls) are offered to the dead and also eaten by the living. In the Miyagi Prefecture area, green soybeans mixed with glutinous rice cakes (zunda mochi) is also eaten at this time.

Finally, rice symbolizes the land of the household which is usually inherited intact by one heir without splitting it up among the other siblings. Land, therefore, is seen as permanent property of a household in which family members constitute a temporary phase.

THE POLITICAL AND ECONOMIC MEANING OF RICE

Besides its religious meaning, rice also signifies wealth and power, being the principal means of taxation during feudal times. The wealth of the feudal lord of each of the 270 domains in the Tokugawa Period (1603-1868) was evaluated through cadastral surveys measuring the rice productivity of the domain land.

In the postwar era the Liberal Democratic Party has benefitted from the voting support of farmers who enjoy the disproportionate voting power and block lobbying power of Nōkyō, the agricultural cooperative. While the farming population only represented 16 percent of the country in 1987, rural votes were worth several times that of their urban counterparts. Rice price subsidies in their various forms have maintained the alliance between farmers and politicians.

As in Tokugawa times, the amount of land rights (both ownership and use rights) in addition to the productive capacity of the land are measures of a farmer's economic relative properity or lack thereof. In a country in which there is a shortage of arable land, land has taken on new meaning to which is assigned economic value. The total acreage in Japan is worth more than that of the United States. Accordingly, the productive value of land and the meaning of the crop produced on it has been thrown awry by the convoluted economics of commercial real estate values. In 1988 the value of the average family farm with a scale between two and three acres was approximately $250,000. The negative side of this paper wealth is that profits from rice production do not justify buying additional land to enlarge the scale of production--especially considering the fact that in 1986 the average farm family earned only 15 percent of their household income from on-farm activities (Nihon Nōgyō Nenkan Kankōkai 1988:536).

JAPANESE RURAL DEVELOPMENT

Japanese rural development has three characteristics which distinguish it from neighboring Korea and Taiwan. These characteristics are: (1) decentralized rural industry and a high rate of part-time farming; (2) a high standard of income parity between rural and urban households; and (3) relative stability in the number of farming households and the scale of agriculture maintained through single heir inheritance.

Huang has described Taiwan as being a case of "agricultural degradation." The dominant features of "agricultural degradation" include: (1) minimal agricultural sectoral growth and inability of the rural population to adapt traditional agriculture to new circumstances; (2) government policy favoring the non-farm population and a growing income disparity between farming and non-farming people; (3) a lack of domestic agricultural development which leads to domestic social problems and increasing dependence on foreign agricultural imports which drain foreign capital reserves (1981:4).

Japanese development differs on each of the following points. First, in Japan there has been steady growth of the agricultural sector and consistent investment in the agricultural infrastructure, particularly land improvement. Land improvement includes projects such as rationalization of paddy size and consolidation of cultivator's plots, improvement of irrigation and drainage facilities, and construction of agricultural roads. Second, government policy has supported agricultural price subsidies and protected certain agricultural products, such as rice, from foreign competition. Government policy has reduced taxes and instituted part-time wage labor opportunities to supplement farm income that have kept farming incomes at a par with urban industrial wages. Third, foreign agricultural purchases have served as an equalizing factor in the balance of trade for Japan's industrial customers. In other words, as a result of decentralized industrial development, an improved livelihood in both agriculture and industry has been provided in a manner not found in Taiwan.

Decentralized Rural Industry and High Rate of Part-time Farming

Compared to most other industrialized countries, a higher proportion of the Japanese population is still engaged in farming, although their income has been increasingly supplemented by off-farm work in industry. Table 1.3 reveals trends in the farming population and number of households in Japan, Korea, and Taiwan. Despite rapid industrialization, as a whole the three countries have maintained a significant number of farmers as a percentage of the total population. In 1980 the agricultural product comprised only 3.8 percent, 16.9 percent, and 9.3 percent of the gross domestic products in Japan, Korea, and Taiwan, respectively. In Japan, agriculture as a percentage of the gross domestic product was about 2 percent in 1988, yet agriculture still employed many people. Because off-farm income opportunities offer a higher monetary reward than on-farm, farming has taken a back seat to industry and, as shown in Chapter 7, the focus has been on how to save time on agricultural tasks to reinvest in industrial wage labor. The increase in the percentage of persons employed by the secondary and tertiary sectors will be explored in depth in Chapter 9.

Thus, the agricultural sector of the economy has been incorporated into, and subordinated by, industrial development. Industry was gradually decentralized to take advantage of the cheaper wages and residential continuity provided by rural social institutions, such as the household (ie) and hamlet (buraku). During the development of Japanese capitalism, agriculture and industry have grown mutually dependent. Kazushi Ohkawa (1982:143) has termed this "sectoral interdependence". This is a sharp break from theories, such as that of Walter Rostow (1951), which describe agricultural development as a "precondition" for creating an agricultural surplus necessary for industrial growth. Instead, Ohkawa's view emphasizes "concurrent" development between the agricultural and industrial sectors of the economy.

I think the concept is also applicable to the postwar situation although in this case agriculture clearly becomes subordinated to industry. Therefore, a more appropriate term is "subordinated sectoral interdependence." The agricultural product

Table 1.3

Farming Population as a Percentage of the Total Population in
Japan, Korea, and Taiwan (1955-1985)

	1950	1960	1970	1980	1985
Japan	45%	37%	27%	18%	16%
Korea	63	58	46	29	22
Taiwan	51	49	40	30	n/a

Sources: Republic of China (1955-1988, 1987), Republic of Korea
1955-1988, Nōrinsuisanshō Keizaikyoku Tōkei Hōkokubu (1955-85).

of Japan has decreased from 26.1 percent in 1950 to a mere 2
percent in 1985. Contrary to the conclusions which might be
drawn from such a low figure, agriculture is an integral part of
Japan's modern industrial structure. The national development
plans, beginning with Kakuei Tanaka's plan to "Remodel the
Archipelago" (Tanaka:1973), successfully decentralized industry to
rural areas by creating an infrastructure of roads and railways.
The dual phenomena of a high degree of industrial subcontracting
and part-time farming go hand in hand. Japanese rural areas
have been transformed by the symbiotic relationship between
agriculture and industry. This is often referred to by those who
favor it as noko ittai or "the harmonious co-development of
agriculture and industry" while detractors have criticized the
exploitation of rural cheap labor.

Following the Meiji Restoration of 1868, Japan set forth on
the path to development, and from the turn of the century to
the end of World War II became the colonizer of Korea and
Taiwan. It is difficult to estimate the total effect of Japanese
policy on its colonies. One thing is certain, however. To the
Japanese, control of food was a principal concern.

In Taiwan, centralized irrigation works, such as the one
constructed in the Chia-nan Plain between 1920 and 1930,
facilitated both rice and sugar cane production. However, as
Chen (1977:178) notes, peasants did not participate in either the
planning, construction, or control of the irrigation system except
as wage laborers. This last point is a major difference between
Japan and Taiwan. Japanese irrigation systems were community
planned, constructed, and controlled as described in this study.
Very often they were a joint effort between the local domain

lords, the ruling elite of local areas, and the peasants themselves. Using land improvement projects in Japan as a model for their colonies, the Japanese also rationalized the shapes of irregular paddy fields in Taiwan and Korea during the occupation. In addition to rice, sugar cane was also promoted in Taiwan because of the warm climate. Increased agricultural production in the Japanese colonies of Korea and Taiwan led to the 1930 rice price tariffs at home to protect Japanese rice producers from the influx of foreign rice.

Table 1.4 compares part-time farming rates in Japan, Korea and Taiwan. Japan and Taiwan are statistically similar with respect to part-time farming. The trend in Taiwan is to commute long distances to the city because urban wages are considerably higher than rural wages. On the other hand, Korea has concentrated on urban industrial development thereby limiting the opportunities for part-time work.

Table 1.4
Part-time Farming Rates in Japan, Korea, and Taiwan

	1950	1960	1970	1980
Japan	50%	66%	84%	87%
Korea	11	9	15	18
Taiwan	--	52	70	90

Sources: Republic of China (1955-1988, 1987), Republic of Korea 1955-1988, Norinsuisansho Keizaikyoku Tokei Hokokubu (1955-85).

Part-time farming rates in Japan and Taiwan do resemble each other, though, when compared to Korea. In Korea the rate of part-time farming remains low due to the government's pursuit of urban industrial development. The Saemaul Movement, Korea's major movement for rural development, focused more on improving rural living conditions than on fostering rural non-farming jobs.

In Japan part-time farmers can be divided into two groups. The first (PT I) are part-time farmers who earn more money on the farm than in outside industry. The second type (PT II) earn more money in outside jobs than in farming. Table 1.5 illustrates how industrial opportunities have favored PT II.

Table 1.5
The Number of Full- and Part-time Farming Households in Japan
(1955-1985)   (X 1000)

Year	Full-time Farming Households	Part-time I Households	Part-time II Households
1955	2,105	2,275	1,663
1960	2,078	2,036	1,942
1965	1,219	2,081	2,365
1970	831	1,802	2,709
1975	616	1,259	3,078
1980	623	1,002	3,036
1985	626	775	2,975

Source: Nōrintōkei Kyōkai 1986:122-123.

Increased Standard of Living and
Income Parity Between Farmers and Workers

Farming households have been able to raise their standard of living and maintain income parity with urban workers. Table 1.6 shows this income parity with farm income being slightly higher.

As a group, farmers have been able to maintain their low debt to savings ratio (about 17 percent) and have actually increased their accumulated savings to yearly income ratio from 82 percent in 1960 to 214 percent in 1980. In 1986 the average farm household had 13,510,000 yen ($112,583) in savings minus debts, which was about three times that of the urban family with 4,418,000 yen ($36,816).   Nōkyō, the agricultural cooperative, plays a key role in this through handling government rice marketing and price subsidies. This will be explored further in Chapter 8.

The income of part-time farmers in Japan has been able to consistently keep pace with and even surpass that of full-time farmers and full-time workers. Efficiently managing two principal resources, land received through single heir inheritance and labor reproduced and received through household alliances, Japanese farming households have created a "part-time farming" survival strategy and have become a vital link in the expansion of industrial capitalism.

Table 1.6
Per Capita Household Income of Japanese Farmers and Workers
(1965-1985)

Year	Per Capita Farm Household Income (X 1000 yen)	Per Capita Worker Household Income (X 1000 yen)
1965	158	194
1970	330	358
1975	870	760
1980	1,273	1,111
1985	1,596	1,421

Source: Nōrintōkei Kyōkai 1986:19.

Townships actively seek out and invite industry to locate in their area to increase the tax base as well as to provide jobs that will raise the standard of living of the farmers. This situation also has been advantageous to industry, which pays lower wages and taxes in rural areas than in the city. Typically, as is shown in Chapter 9, the size of the firm attracted to rural areas is small. In Japan the rate of subcontracting is much higher than in the United States. In 1985, 55 percent of all manufacturing employees in Japan worked in firms with fewer than one hundred employees. In the United States this number was 28 percent in 1982 (Chūko Kigyōchō 1987).

Farmers in Japan enjoy a much higher standard of living now than in the immediate postwar period. They have, however, slipped into a cycle of dependency on wage labor income to purchase farm machinery. Many who start out with a side job to merely supplement their income, end up buying labor-saving farm machinery. This allows them to increase off-farm income. This extra income is needed to pay for the machinery. Table 1.7 demonstrates how on-farm income has been decreased as a result of off-farm wage labor in industry.

Relative Stability in the Number of Farming Households

Japan, Korea and Taiwan vary in inheritance patterns and household organizations. The institution of single heir inheritance

Table 1.7
On-Farm Income Compared to Total Income for Japanese Farm
Households (1955-1985)

Year	On-Farm Income (A) (X 1000 yen)	Total Income (B) (X 1000 yen)	(A/B)
1960	225	449	50.2%
1965	365	835	43.7
1970	508	1,592	31.8
1975	1,146	3,960	28.9
1980	952	5,594	17.0
1985	1,066	6,915	15.4

Source: Nōrintōkei Kyōkai 1986:23.

in Japan contrasts with Korea where the eldest son inherits most
of the land while lesser amounts are allocated to other sons
(Brandt 1971:113; Lee 1976:12; Sorensen 1988:219). In Taiwan
sons ideally receive more or less equal shares of the land
inheritance (Freedman 1965:23-25; Wolf 1972:167).

The inheritance pattern has a direct bearing on the number
of farming households over time. Table 1.8 again compares
Japan, Korea, and Taiwan and shows the divergent pattern of
postwar farming in East Asia. At one end of the continuum is
Japan with a 75.5 percent decrease in the number of farming
households, while Taiwan is at the other end of the continnum
with an increase in number of farming households. Korea is in
the middle.

No doubt the Taiwanese practice of equal son inheritance
has fragmented land holdings and consequently decreased rural
earning capacity. According to the Taiwan Agricultural Census,
the percentage of farms less than 0.5 hectares in area increased
from 34.4 percent in 1955 to 41.7 percent in 1975. Part of this
fragmentation is due to population increase coupled with the
multiple heir inheritance of land.

If inheritance was single heir instead of multiple heir, the
number of households in Taiwan should have decreased. This
might have contributed to making it possible for Taiwanese rural
farmers to enjoy an income comparable to urban workers. Huang
(1982) puts the blame on the national policy favoring urban
workers. While this is no doubt a factor, it is equally true that

Table 1.8
The Number of Farming Households in Japan, Korea, and Taiwan (1950-1980)

Year	Japan	Korea	Taiwan
1950	6,176,419	2,233,562	638,062
1960	6,056,630	2,349,506	800,835
1970	5,341,800	2,483,318	879,005
1980	4,661,384	2,155,915	872,267
1980/1950	75.5%	96.5%	136.7%

Sources: Peng (1987), Republic of China (1955-1988), Republic of Korea 1955-1988, Nōrinsuisanshō Keizaikyoku (1955-85).

Taiwanese equal male inheritance pattern bears responsibility for the land fragmentation.

This book demonstrates the structural adaptability and persistence of the Japanese household during the development of industrial capitalism. Chapters 4, 5, and 6 argue that the inter-generational maintenance of shared rights over land and intra-generational flexibility of the composition and allocation of its labor force have enabled the household to persist and thrive as a domestic corporate group.

This study presents national statistics whenever possible and uses the example of a community located in the heart of the rice belt. Based on participant observation, the study focuses on Nakada Township in northern Miyagi Prefecture, which is one of the six prefectures that comprise the Tōhoku Region. The Tōhoku Region leads the country in the amount of rice produced and productivity. Nakada Township in Tome County was chosen for its high rice productivity (third in Miyagi Prefecture), high rice quality (96 percent grade 1), and excellent tasting Sasanishiki variety of rice. The land has a slope suitable to large-scale farming but was in 1988 still converting one-quarter acre paddies into three-quarter acre paddies. The average land farmed by Nakada farmers is 1.3 hectares (3.2 acres) or just over the national average and, having attracted several large factories and numerous subcontracting companies, part-time farming rates are likewise close to the national average. Table 1.9 gives the national, Tōhoku Region, and Nakada Township rates for full and part-time farming.

Table 1.9
<u>Full- and Part-time Farming Averages for Japan, Miyagi
Prefecture, and Nakada Township in 1985</u>

	Japan	Tōhoku Region	Nakada Township
Full-time Farming	14.3%	9.3%	9.1%
Part-time Farming			
Type I	17.7	27.0	29.3
Type II	68.0	63.7	61.6

Source: Nōrinsuisanshō Keizaikyoku Tōkei Hōkokubu 1985:331.

NOTES

1. See Holtom (1972) and Ellwood (1973) for a full description of the Japanese enthronement and accession ceremonies.

2

The Environment and Settlement Pattern of Nakada Township

ENVIRONMENT

Sandwiched between the Kitakami River in the east (the largest river in Tōhoku Japan), and the Natsugawa River in the west, the settlement pattern of Nakada Township (38 degrees 42 seconds north latitude and 141 degrees 15 seconds east longitude) reflects the historical development of irrigation lines, various agricultural land improvements, and the continuing struggle between the people and the major river of northern Japan. The Kitakami River is fed by the melting snow and rainfall from the mountains of Iwate Prefecture one hundred miles to the north. For this reason, the water flow is unsteady and unpredictable and has been a cause of constant concern for the inhabitants of Nakada Township, who depend on the water for wet rice irrigation. The headwaters of the Natsugawa River are in the Kurikoma Mountains to the northwest. These rivers form natural boundaries with the surrounding townships of Tōwa, Ishikoshi, and Hasama. Flooding from both rivers, especially the Kitakami, has always been a problem although the silt brought by such disasters no doubt helped increase the fertility of the soil and this region is famous for delicious rice. The location of Nakada Township, Tome County, Miyagi Prefecture is shown on Figure 2.1.

Nakada Marsh, from which the township derives its name, lies in the northern section of Nakada Township and borders on Hanaizumi Township in Iwate Prefecture. Nakada Marsh was a natural marsh formed by drainage from the mountains in Hanaizumi and the floodwater of the Kitakami River. The marsh was reclaimed in 1908 to form new paddy fields.

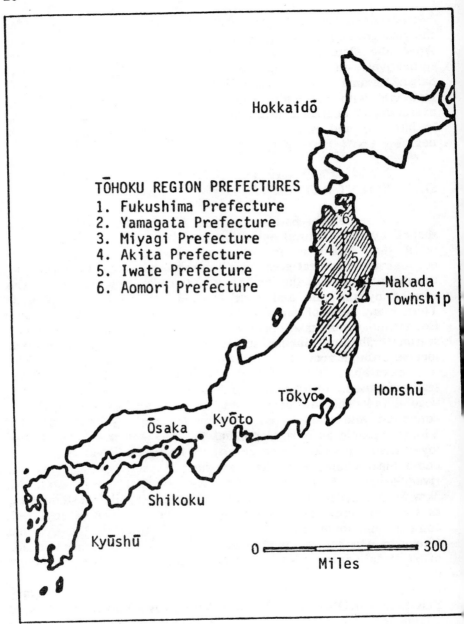

Figure 2.1
Location of Nakada Township, Tome County, Miyagi Prefecture

Between 1965 and 1980, the average monthly rainfall during the rice growing season ranged from a low of 72.7 millimeters in April and October to highs of 137, 124, and 138 millimeters, respectively, in July, August, and September (Miyagi-ken Kisho Saigai Hasama Chihō Taisaku Kaigi 1981). The temperature is moderate with the temperature during early May rice planting averaging 17 degrees centigrade (63 degrees F.) and summer highs in late July and early August reaching 29 degree centigrade (84 degree F.).

SETTLEMENT PATTERN

The settlement pattern of Japanese villages is in large part shaped by their natural environment. The reader should bear in mind that Japan is an archipelago with a mountain chain occupying the central area from top to bottom. On both sides of the mountain chain, the land slopes rapidly to the sea, and rivers formed by the rain and melting snow are swift and seasonal. Thus, many areas of Japan suffered from the combined effects of floods and droughts. For this reason, areas which experienced seasonal flooding during spring rains or typhoons often are located on high ground.

Sakuraba Village ("Cherry Tree Village") upon which this study focused, utilizes the high ground for the growing of vegetables. The low ground was formerly marsh and later was converted into paddy land. The hamlets comprising Sakuraba Village encircle Nakada Marsh, which even now is several meters lower than the surrounding ground. The first settlers in the area chose the highest fertile land leaving the lower land next to the river or in the flood zone to the less fortunate. Nakada Township's settlement pattern, in which hamlets contain as many as sixty to eighty households, contrasts with its major rice belt competitor, Shonai Plains. Shōnai Plains is located in neighboring Yamagata Prefecture on the Sea of Japan and its small hamlets often contain fewer than fifty households.

FOUR ENVIRONMENTAL ZONES OF CHERRY TREE VILLAGE

The four environmental zones of Cherry Tree Village are defined by Nakada Township's proximity to the Kitakami River and the fact that Nakada Marsh traditionally served as a reservoir for the surrounding area. These environmental zones necessary for understanding the lives of the people of Nakada Township are shown in Figure 2.2. These environmental zones are: Zone One or Nakada Marsh; Zone Two or The Traditional Rice Fields; Zone Three or The Traditional Vegetable Fields; and Zone Four or The Fields Next to the River. The ecological characteristics of these four environmental zones are given below.

The first environmental zone, Nakada Marsh (Nakada Numa), is a peat based marsh which was reclaimed into rice paddies. Although the marsh was drained in 1908, it has only been in recent years that the soil has been improved to the point that top grade rice can be grown. Before the 1950s the soil was mucky and difficult to work. Sometimes the farmers would sink thigh deep in the soil during the spring planting because the natural drainage was poor. Since the early 1950s, however, the soil has been improved by bringing in "guest-soil" (kyakudo). "Guest soil" refers to soil brought in from neighboring hillsides and mountains to enrich the soil fertility.

At a depth of about sixteen centimeters, Zone One contains a peat layer beneath the kyakudo. This peat layer, while rich in nitrogen, can cause plants to die from root rot when too much hydrogen sulfide is generated during the warm summer months. Also, during a flood in 1949 one of the rice paddies in the deepest part of the former marsh floated as a result of its buoyancy. At approximately fifty centimeters, the soil turns from ash-colored to black.

Zone Two, the traditional rice fields (kiseiden), surround the marsh. The earliest documentation[1] of the marsh shows that this area was used for rice cultivation from 1720, but probably rice was also grown here before this date. The soil is bluish ash which turns black at between eleven and thirty-eight centimeters. A bottom peat layer can be encountered under the black layer, indicating that the marsh was probably larger prior to human efforts to contain the river flooding which feeds the marsh. The soil of Zone Two is well drained and contains a significant amount of silt deposited by floods.

The third environmental zone, the traditional vegetable fields (hatake), is at the highest elevation. Flood waters seldom reach this zone. In the lower areas there is some sand and there are red streaks of iron mixed into the brownish soil. Vegetables, grains, and fruits were grown here because the soil conditions and higher elevation favor these crops. In areas of this zone which were subject to a rare flood, mulberry orchards

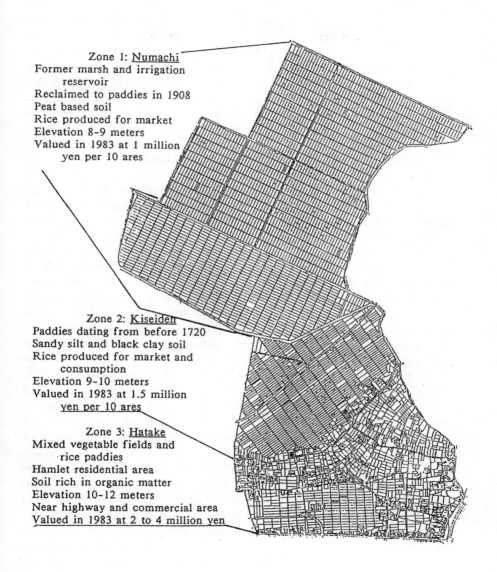

Zone 1: <u>Numachi</u>
Former marsh and irrigation
    reservoir
Reclaimed to paddies in 1908
Peat based soil
Rice produced for market
Elevation 8-9 meters
Valued in 1983 at 1 million
    yen per 10 ares

Zone 2: <u>Kiseiden</u>
Paddies dating from before 1720
Sandy silt and black clay soil
Rice produced for market and
    consumption
Elevation 9-10 meters
Valued in 1983 at 1.5 million
    <u>yen per 10 ares</u>

Zone 3: <u>Hatake</u>
Mixed vegetable fields and
    rice paddies
Hamlet residential area
Soil rich in organic matter
Elevation 10-12 meters
Near highway and commercial area
<u>Valued in 1983 at 2 to 4 million yen</u>

Figure 2.2
Environmental Zones of K Hamlet in Cherry Tree Village

were planted because they were hardy enough to withstand flooding. In other areas soybeans, wheat, barley and everyday vegetables for home use were planted.

The fourth and final environmental zone is comprised of the fields next to the river (kawabatake). These loam soils have the highest content of sand and are ideal for many vegetables. However, risk of flooding is greatest in Zone Four because the zone is between the river and the dike. Like Zone Three, mulberry trees were common in this zone because they could withstand the more frequent flooding.

SETTLEMENT PATTERN OF CHERRY TREE VILLAGE

Cherry Tree Village is one of five villages that historically surrounded Nakada Marsh. As is the usual pattern in rural Japan, the houses in the village are clustered in groups of fifty to one hundred households called hamlets (buraku). The hamlets are located in Zone Three described above and therefore the farmers live separated from their paddies, which are located in several of the other environmental zones. K Hamlet, one of six hamlets comprising Cherry Tree Village, is the focal point of this study and contains seventy-two households, fifty-eight of which were farm households.

In 1885 Cherry Tree Village was amalgamated into Uwanuma Village which in turn was amalgamated with three other villages to form Nakada Township in 1956. According to local population registers such as the fūdoki (Nakadachōshi Henshū Iinkai 1977), in 1772 the population of Cherry Tree Village was fifteen hundred people in 205 households. Thus, there were approximately 7.3 persons per household. This compares to only 5.1 persons per household in downstream Mizukoshi ("Water Runs Over") Village, which had a population of 1,263 persons in 250 households. Mizukoshi Village, through which the Kitakami River flowed until the river was rerouted in the early seventeenth century, possessed fewer rice paddies because of frequent flooding and a poor irrigation system. Cherry Tree Village, on the other hand, enjoyed a commanding position on the irrigation system and an abundant water supply. It is possible that the large difference in the number of persons per household between Cherry Tree Village and its neighboring downstream village was possibly related to the higher intensity of labor required by the greater proportion of wet rice land. Demonstrating a clear difference in wealth,

Cherry Tree Village also owned more draft animals than its neighboring village. In Cherry Tree Village there were a total of 120 horses compared with only sixty-eight in Water Runs Over Village.

HISTORICAL DOMAIN CONTROL

One of the earliest historical accounts of Cherry Tree Village dates to the remains of Cherry Tree Castle which was constructed in 1338 and was closely connected to the Kasai Family. The area is characterized by many changes in domain affiliations. In 1097 it was part of the Fujiwara Domain of Hiraizumi and thereafter was alternately controlled by the Fujiwara and the Kasai family domains.

A series of small forts dating between 1300 and 1600 were built in Nakada Township. The Kasai Family ruled the area during the sixteenth century until defeat by the Date Domain in 1604. Following their victory, the Date Domain built the first dikes along the Kitakami River, making possible the opening of new rice fields.

The Date (Sendai) Domain forbade divisions of land through inheritance (bunkatsu sōzoku) and put upper and lower limits on how much land a household could till (Tamayama 1948). The idea behind the rule was to maintain the rice tax flow from the peasants and avoid overpopulation which would consume too much rice. The policy was established between 1719 and 1728 and was maintained until the Meiji Restoration of 1868. The mandate set an upper limit on tilled land of five kanmon[2] and a lower limit at five hundred mon (later to be reduced to three hundred mon), which amounted to a range between 0.6 acres and 9.6 acres depending on the yield of the land. Lord Date encouraged peasants to increase rice production by permitting them to exceed the five kanmon limit if they developed new land (shinden). The labor invested in reclaiming and developing farm land was rewarded with a tax waiver for ten years. According to the land cadastre of 1800 for Cherry Tree Village, the shinden approach was common.

DESCENT PATTERN OF CHERRY TREE VILLAGE

Primogeniture has continued to be standard inheritance practice in Cherry Tree Village since recorded history. The type of primogeniture shifted around the turn of the century from first child of either sex (ane katoku)[3] towards favoring the first born male child as heir. In both cases the custom of single heir inheritance has been the fundamental principle insuring perpetuation of the household (ie). Despite the change in the inheritance pattern, families in Nakada Township still celebrate the birth of their first child, regardless of sex, in a formal ceremony called "grandchild celebration (magoburumai)." The ceremony is usually not repeated for subsequent siblings.

During the last part of the nineteenth century up through the end of World War II the eldest male child usually inherited. After the war, the United States occupation imposed a new Civil Code with equal inheritance laws, but these never really took hold in rural areas. Social norms required that non-inheriting siblings relinquish their legal rights so that the household would be able to continue intact without fragmenting its land holdings. Non-heirs became spouses in other households and migrated to urban areas providing the labor supply for industry.

Numerous scholars (Ariga 1943; Bachnik 1981; Fukutake 1980, 1981; Moore 1985; Nakane 1967; and Yanagita 1951) have drawn attention to the importance of the resource base to the continuity of the Japanese household. Some, such as Fukutake 1980:54-63), feel that with the growth of contractual farming, the ie system will gradually give way to equal inheritance.

The Nakada Township evidence does not support this last contention. Despite rapid industrialization, farmers have tenaciously held on to their land rights and the number of farm households has been maintained fairly constant since the war. Presently farm households are faced with a number of problems such as a bride shortage and loss of jobs and agricultural subsidies, described in Chapters 9 and 12. The ie is a flexible unit, however, and may be able to shift its resources into non-farming areas. In other words, the type of resource base can change but this change does not in itself bring about a structural reorganization of the unit. Recent studies (for example, Kunihiro 1984 and Izumi et al. 1984), of both urban and rural ie dwellers, indicate that nearly three-quarters prefer their eldest son to inherit the house and that efforts are made to side-step the equal inheritance law (Otohiko 1985:9).

Cherry Tree Village can also be characterized by a high rate of household branching during Japan's feudal period. The main household (honke), and its branch (bunke) households and their subsequent branches form large related clusters or groups called edōshi[4], or ikke mage in the local dialect. In the village each of the three largest clusters presently contain between twelve and fifteen households. Most edōshi clusters, however, have fewer than ten households[5].

The formation of these clusters depended on a balance between land and labor resources. For example, in K Hamlet, where one of the three largest Cherry Tree Village edōshi is located, one edōshi now has ten branches in addition to the main household. According to a cadastre, in 1800 there were only two branches in this edōshi in addition to the main household. During the nineteenth century the edōshi formed two more branches, with the remaining six households fissioning between 1902 and 1939. In the same hamlet a different edōshi group of twelve households was formed sometime during the four hundred years preceding the 1800 cadastre. In both the above cases, there was a considerable difference between the amount of land held by the main household and that held by its branches.

There seems to have been several methods by which households branched. These methods demonstrate that both resources and a moral system of reciprocal rights and obligations are used to maintain the system. According to my analysis of the household registers (koseki) for Cherry Tree Village, the highest frequency of branching occurred secondary to efforts by the household to resolve problems of genealogical succession. For example, after the heir married but did not produce offspring, to reduce the risk of an heirless household, a marriage of the heir's sibling was arranged. Then, unexpectedly, both couples would produce children. The rule of single heir inheritance required that only one child succeed so as compensation one of the families was set up as a branch household. Demonstrating that Japanese kinship is not limited to blood relationships, branch households were sometimes established using a non-related family who had faithfully worked for the main household. In Cherry Tree Village, the usual practice was to allocate a small parcel of land to the branch after the branch household tilled the land for several years. The branch members were obliged to work for the main household and the relationship in many cases resembled that of landlord and tenant. In fact, there were many such cases that surfaced during the land reform.

In some of these a branch obtained ownership rights to land that it had tilled for its main household. In other cases, the tenant branch relinquished its right in order to maintain harmonious relations with its main household.

In the Confucian tradition it was also expected that if the branch had problems, the main household would be benevolent. If the problem was financial, rents would be lowered in years of poor harvest. If there was a problem with genealogical continuity, it was not uncommon for the edōshi main household to give one of its offspring or an elderly person as the heir to its branch.

As Keith Brown (1968:115) has pointed out, the relationship has a content that "is characterized by a value or ideal for a mutual solidarity which is functionally diffuse and is flexible in its manifestations and expressions." Edōshi in Nakada Township usually get together for funerals, marriages, and New Years Day, during which time they sit in rank order according to branching chronology. They also usually share a household taboo (karei) such as prohibiting the planting of a particular crop. The taboo "Don't eat fish or eggs during rice transplanting" was a taboo held in common by the largest edōshi but was stopped in the 1960s when the edōshi collectively hired a priest to exorcise the spirits. Karei almost always relate to the agricultural planting cycle and, being regional in occurrence, have not been reported in other literature about ie or dōzoku.

SHARED LAND IDENTITY

Because the branch households receive land from their main households, it is common for main and branch households to own land adjacent to each other. Figure 2.3 illustrates the grouping of home sites of one of the edōshi in K Hamlet. Figure 2.4, based on a reconstruction of the irregular traditional paddies, shows land holdings in the Zone Two environmental zone, or the traditional rice fields. The edōshi was able to keep contiguity when the paddies were reorganized into rectangular paddies in 1955.

Households in Cherry Tree Hamlet demonstrate a shared identity with their land and maintain it through the practice of cooperative drainage ditch cleaning (eharai). A representative from each farming household in Cherry Tree Hamlet pitches in on an equal basis (without regard to size of land holdings). This

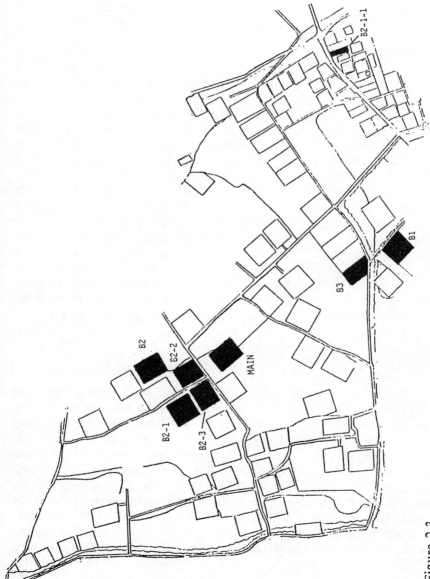

Figure 2.3
Edōshi Settlement Pattern in K Hamlet

30

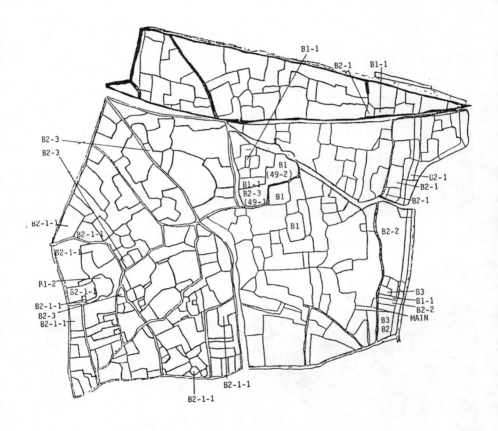

Note: "B2-1-1" refers to the first branch household of the first branch household of the second household to form a branch from the main household.
Source: Nakada-chō (1892, 1983), koseki[6].

Figure 2.4
Edōshi Land Holdings in a Traditional Paddy District

Photo 2.1
Magoburumai Celebration for the First Grandchild

Photo 2.2
Semi-annual Cleaning of Rice Paddy Drainage Ditches by Representatives of Each Farm Family in K Hamlet

rite, which occurs in early spring and mid-summer, demarcates the irrigation rights and property boundaries of hamlet members and serves to draw attention the their shared identity.

Another example of shared identity exists at the village level. At one time a forest, which was held as a commons, was within the boundaries of Cherry Tree Village. As land became more valuable, it was leveled and used for paddy land, the proceeds of which are still communally owned.

ASSOCIATIONS

There are two major associations in K Hamlet. One of them is the mutual aid association (keiyaku kō). It is comprised mainly of landowning households and is organized according to neighborhood groups. Within each association there are sub-groups (kumi). The five mutual aid associations in K Hamlet are shown in Figure 2.5. I have also shown the kumi groupings (A-D) for Mutual Aid Association Number Four. Each household is associated with one of the kumi. Mutual aid associations are neighborhood specific and it is possible for branch households of the same edōshi to belong to different mutual aid associations. As described earlier, households are affiliated with the mutual aid associations of their respective neighborhoods. The associations meet twice a year, once in the spring and once in the fall, for the purpose of mutual-aid and for the duties relating to funerals. The meeting place and the records of their meetings (shoruibako) is rotated between members of each kumi and then rotated from that kumi to the next kumi in line. The kumi next to the household where a person dies notifies relatives and friends about the funeral and serves the funeral meal. For instance, when a person from A Kumi dies, the households in B Kumi would serve the funeral meal. The associations sometimes plan trips to use up surplus funds generated by the interest from the association bank account into which they put their annual dues.

The other major association is the Women's Fertility and Household Well-Being Association (Yama no Kami Kō). Like the mutual aid associations, the Yama No Kami Associations meet twice a year. There are three such associations in K Hamlet. The membership of the Yama No Kami Association is more on a voluntary basis than the Mutual Aid Association although households are seldom invited to join unless they have

Photo 2.3
Mutual Aid Association Meeting in K Hamlet

Photo 2.4
Edōshi Branch Household Funeral With Household Representatives
from Main and Branch Households Seated in Rank Order

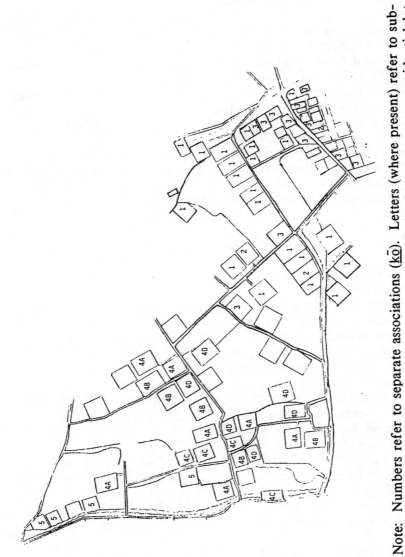

Note: Numbers refer to separate associations (kō). Letters (where present) refer to sub-association groups (kumi). Each rectangular box represents a residential lot.

Figure 2.5
K Hamlet Mutual Aid Associations

demonstrated an intent to live permanently in the hamlet. Since young brides are the most active members of this popular association, some households without young brides do not attend. They refer to this as "taking time off" (yasundeiru) since it is their intention to have their future young bride attend on a regular basis. At meetings members bless rice in front of the Yama No Kami rock to which they pray. Each member drinks rice wine and takes a portion of the newly sanctified rice home to mix with the rice they cook for their families.

In summary, the basic building block of the K Hamlet social structure is the land owning ie. The main purpose of this chapter has been to demonstrate that rights over land constitute a necessary condition for effective household operation, branching, and even participation in the mutual aid associations. With rapid industrialization and the declining importance of land as the fundamental resource of the ie, it is possible that in the future the importance of rights over land will be somewhat diminished.

NOTES

1. The oldest document dates to Kyoho 5 (1720) and shows the legal village boundaries that divide the marsh area. It is hand drawn on rice paper with the official signatures of village leaders and is housed in the home of Hajime Onodera of Ishinomori, Nakada Township.

2. Most domains in Japan used the term "kan" with the kokudaka system. The kokudaka system was based on the total amount of crop yield in one region. In Ōsaka and Kyōto one kan equalled five koku but in the Date Domain it equalled ten koku of hulled unpolished genmai brown rice. One koku, a measurement based on the volume of a woven straw bag of rice, was approximately 0.18 cubic meters and weighed slightly less than 142 kilograms (302 pounds). One thousand mon equalled one kanmon. In the Date Domain, the term "kanmon" referred to the amount of land which should produce a certain amount of rice. Therefore, in the Date Domain, one kanmon denotes a different amount of rice depending on the quality of the land (Tamayama 1948).

3. Suenari (1972a and 1972b) has described first child inheritance in the Tōhoku Region.

4. Edoshi is the local term for what many anthropologists and sociologists have termed dōzoku. As Izumi et al. (1984) and

Nagashima (1984) have shown, significant regional variation exists with respect to kinship types in Japan. Edōshi tend to be hierarchical.

5. Cornell (1963) has demonstrated the stability of rural residential groups.

6. The Ministry of Justice granted official copies of the household registers (koseki) for six hamlets to me. These were used in conjunction with land register data as background information for the personal interviews.

3

The Development of Irrigation

Mizu (Water)

Moving on its own accord, water causes other things
  to move;
Always seeking its own path, water cannot be stopped;
Meeting obstacles, water doubles its force,
Being in itself pure, water cleanses impurity from
  others;
And is the sum of joining impure with pure.

By Ryūsensho, Chief Abbot of the Sōdō Zen Sect

Source: Displayed in the home of the mayor of Nakada
  Township[1].

THE SIGNIFICANCE OF IRRIGATION
TO JAPANESE AGRICULTURE

Wet rice needs a constant supply of water to insure healthy
and timely development.   Water temperature and rate of flow
through the paddy also affect the crop.

There have been four stages of irrigation development in
Nakada Township.   Each stage brought about a new relationship
between individuals, between villages, and with the natural
environment.   This chapter explains the necessity of examining
inter-village social relations with respect to control over the
irrigation system.   The relative control or lack of control over

water has played and continues to play an important role in land tenure relations.

The irrigation system of Nakada Township is a variation of "old pond irrigation" systems as described by Nagata (1971, 1979). Pond irrigation systems typically are fed by hill or mountain runoff and/or diversion from rivers. According to Nagata, Japanese irrigation systems typically have upstream-downstream disputes over water. These disputes revolve around upstream efforts to maintain control over irrigation water and downstream efforts to gain a stable supply of water[2].

The first stage in the four stages of irrigation development in Nakada Township was the use of Nakada Marsh as a reservoir for irrigating the surrounding villages. Stage Two was the reclamation of Nakada Marsh and the establishment of the upstream Ōizumi Pump Station located next to the Kitakami River. Stage Three was the postwar creation of the Land Improvement Office (Tochi Kairyō Ku), which became responsible for water regulation. The Land Improvement Project (Tochi Kairyō Jigyō), which fostered the placement of the Asamizu Pump Station downstream making the distribution of water more equitable, was Stage Four.

STAGE ONE: NAKADA MARSH AS A RESERVOIR

Nakada Marsh, until it was reclaimed in 1908, covered approximately 541 hectares and served as a reservoir for rice irrigation while producing much of the fodder needed for draft horses. A smaller marsh adjacent to Nakada Marsh was called "Little Nakada Marsh" and was approximately forty-one hectares in size. These marshes supported carp (funa), catfish (namasu), eels (unagi), loaches (dojō), freshwater shrimp Metapenaeus burkenroadi (moebi), and shellfish (kai), all of which were eaten by the locals. Numerous birds, such as the Siberian white swans (hakuchō), migrated to the marshes in the winter and provided yet another food source. Various grasses were harvested from the edges of the marsh along with water chestnuts (hi shi no mi), lotus flower and lotus roots (renge no hana), water lilies Nymphaea tetragona, water caltrop (hitsujigusa), curly pondweed (ebi mo), bladderwort Lentibulariaceae (tanukimo), and watershield Brasenia purpurea (junsai).

The marsh supplied irrigation water by gravity flow. A gravity flow tunnel dug through the dike along the Kitakami River during a drought in the 1750s, served as a sluice which let the water into the marsh whenever the river flooded above its normal level. At its lowest level the marsh was about seven meters above the Kitakami River, so the tunnel had to be dug well above the water level of the river. This made it impossible to obtain water during dry spells or when the spring rains were steady and did not provide a river swell. The water level of the reservoir was usually highest after May when the melting mountain snow to the north and the spring rains combined to swell the Kitakami River[3]. Waiting for a river swell often delayed rice planting. During the Tokugawa Period the reservoir directly or indirectly serviced between fifteen hundred and two thousand hectares in the following villages: Uwanuma, Ishinomori, Takarae, Mori, Mizukoshi, Asabe, Yoshida, Yoneyama, Toyori, Sanuma, and Tome. Owing to droughts and the uncertainty of the spring rains needed to fill the reservoir, downstream irrigation in about half the village was very unreliable.

Downstream hamlets such as Mori and Shinden in Takarae Village, which were located on the lower end of the irrigation system, obtained the poorest harvests due to lack of water during droughts, or flooding when the dike broke during large river swells. In 1888 these two hamlets were able to irrigate four hundred hectares by building a canal from the Hasama River. This left the Nakada Marsh Reservoir directly servicing 1,139 hectares listed below in Table 3.1 and also shown on Figure 3.1.

Table 3.1
Nakada Marsh Irrigation Service Areas

Nakada Marsh Reservoir	Service Area
Uwanuma Village	190 hectares
Ishinomori Village	514 hectares
Takarae Village	81 hectares
Asamizu Village	0 hectares
Total	785 hectares

Little Nakada Marsh Reservoir	Service Area
Uwanuma Village	73 hectares
Asamizu Village	280 hectares
Total	353 hectares

Source: Nakada Chōshi Henshū Iinkai 1977: 206.

40

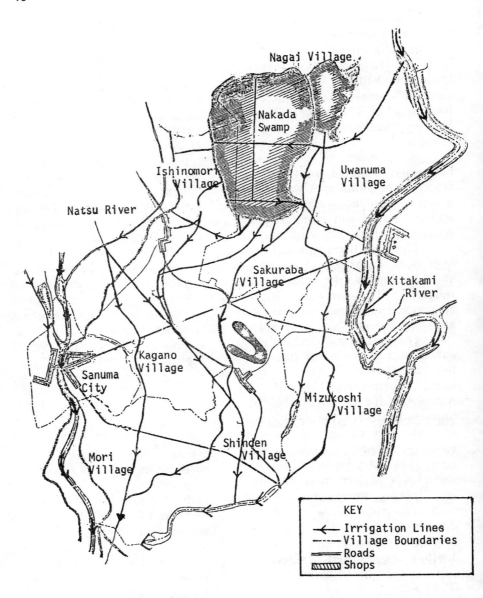

Source: Adapted from Nakadachōshi Henshū Iinkai 1970.

Figure 3.1
Nakada Marsh Reservoir Service Area

Uwanuma Village traditionally controlled the irrigation system, using it to irrigate only 263 hectares of its own land. Ishinomori Village had minor control of the marsh but needed the water to irrigate 512 hectares, about twice the amount cultivated by Uwanuma Village. Asamizu Village was at the mercy of Uwanuma since it received irrigation water from Little Nakada Marsh Reservoir in Uwanuma Village to irrigate 280 hectares. Takarae Village received water for a mere eighty-one hectares from Uwanuma.

As a result of their subordinate positions on the irrigation hierarchy, Takarae and Asamizu suffered economically and often made concessions to Uwanuma and Ishinomori. This situation was partially relieved when Takarae was able to obtain water from the Hasama River and when Asamizu built a pump station next to the Kitakami River following the postwar Land Reform.

The inability to procure water led to absentee landlord control over both Takarae and Asamizu prior to the Land Reform of 1946 and has been the most important factor shaping intervillage relations. The postwar ability of Takarae and Asamizu to acquire adequate irrigation water is one of the most notable changes that has occurred in Nakada Township village social relations.

The reluctance of Uwanuma and Ishinomori Villages to participate in the Land Improvement Project is no doubt related to the fact that their existing irrigation facilities are superior. The Land Improvement Project, described in Chapter 10, presented Asamizu with its own water pump station, giving Asamizu water independence for the first time in its history. As we shall see in the last part of this chapter, the Land Improvement Project has turned a dream into reality for Takarae and Asamizu because they now have the opportunity to surpass Ishinomori and Uwanuma in the quality of irrigation and drainage facilities.

During the seventeenth century there were nineteen floods recorded. Water pouring in through breaks in the dikes brought disaster to most of the township since Nakada Township is only seven to twelve meters above the river level. In the eighteenth century the number of floods increased to thirty-five and then to thirty-six in the nineteenth century. Due to damming projects and flood control measures the number of floods during this century has decreased to ten.

During the most recent flood on July 20, 1947 over 949 of the approximately four thousand hectares of arable land were covered by water. The flood killed six people, destroyed twenty-four houses, and leveled one hamlet located next to the river. Water rose above the floors in over 378 houses. This, however, was not nearly as severe as the 1875 break downstream which killed forty-one people.

Floods were the number one cause of crop failures with drought being the number two cause, both being related to the control and delivery of water. Together these two factors are responsible for more crop failures than early frost, insects (particularly unka leaf hoppers),or disease, such as rice blast disease (imochibyo).

Floods, droughts, pestilence, and plant disease caused great famines (daikikin)[4]. These famines were often accompanied by plague, cholera, measles and smallpox epidemics. During the Temmei Famine of 1783-84, twenty-six hundred people died in Tome County, of which Nakada comprises one of eight townships. During the famous 1833 Tempo Famine, the loss of rice was so serious that many starved and some even ate human flesh. There is a picture in the Tome Gun Shi of a scroll from this period depicting a starving outcast digging for roots on the edge of Nakada Marsh.

The people downstream were hardest hit by floods, droughts, crop failures, and disease. Particularly hard hit were areas such as Mizukoshi Village which the river cut in half before being rerouted in the seventeenth century by the Date Domain.

Many stories are centered around the rivers, the floods, and famines which they caused. There is a tale about a self-sacrificing Mizukoshi man named Oikawa Seitaro who fed barnyard grass gruel (hiekayu) to the starving village peasants during the great famines. Other stories from Mizukoshi include the story of Otsuru Myojin Shrine, dedicated to an innocent servant who was buried alive in the dike near Kawazura Hamlet by order of the second supervisor of the Date Domain in the 1630s. It was believed that this human sacrifice would protect the dike from breaking during floods. Annually local people still offer rice to the Otsuru Shrine located on the inside of the dike.

STAGE TWO: THE RECLAMATION OF NAKADA MARSH AND THE ESTABLISHMENT OF THE OIZUMI PUMP STATION

The movement to reclaim Nakada Marsh was led by people in Uwanuma and Sakuraba Villages who owned or cultivated paddy land. It was opposed by peasants in Ishinomori and Asamizu who potentially would have been cut off from the irrigation system if the marsh was reclaimed. Reclamation efforts were also opposed by fodder-grass cutters at least as early as 1889, at which time the grass cutters destroyed irrigation canals to some paddies being reclaimed from the marsh grasslands. These paddies were probably the ones reclaimed in the early Tokugawa Period but abandoned during the drought of the 1750s.

Because most of the marsh (371 of a total 461 hectares) was inside Uwanuma Village, Uwanuma in 1898 initiated negotiations with an engineering professor at Tōhoku Imperial University. The professor was to receive seventy hectares of reclaimed land plus part of the yield in exchange for paying half the cost of the project (including the cost of the pump). Under this plan Ishinomori would have received 514 irrigated hectares while Takarae would have secured enough water to irrigate a mere eighty-one hectares. Uwanuma, maintaining control of the water, would have maintained its 190 irrigated hectares. With one exception this plan was very similar to the final plan approved. The exception was that this first plan denied water rights to the other villages.

Village usufruct rights over irrigation water were a major problem in the Nakada Marsh reclamation. The marsh was legally reclassified as publicly owned land (kanyūchi) governed by the county administration so other villages were entitled to claim their usufruct rights to the irrigation water it produced. Even though Uwanuma possessed the largest share of the marsh, the other villages held usufruct rights to the irrigation water. As a result, Uwanuma could not legally contract for the project by itself and was forced to obtain the consent of all the other villages which had water rights. Therefore, the county sponsored the reclamation project after the other villages of Sakuraba, Ishinomori, Takarae, and Nagai gave approval. In order to gain approval it was necessary to guarantee delivery of irrigation water to all these areas.

In 1908, with the permission of all the villages, the county undertook the project to reclaim the marsh. The resulting plots were leased out by the county which planned to use the rental income to pay for the land reclamation over an eighteen year period. Thereafter, the rent would generate income for the county.

The reclamation was one of the most profitable in Miyagi Prefecture. Due to the reclamation rental income, each year after 1916 the county was able to divert nearly ten thousand yen to other county development projects. These included the construction of a school dormitory for the Sanuma Middle School, the establishment of a hospital in Sanuma, a car for the Sanuma police department, the construction of the Maya Bridge over the Kitakami

River, hiring of agricultural extension agents, and the establishment of railroad facilities. In Miyagi Prefecture, Tome County is the only county to sponsor a land reclamation project.

In land reclamation projects landlords would usually form a joint venture cooperative to invest their capital. In this way, they directly benefitted from the ensuing increase in crop yields. These joint ventures often formed land improvement cooperatives (kōchi seiri kumiai). Such is the case of the Shinai Marsh Land Reclamation Project, the largest land reclamation project in Miyagi Prefecture. It encompassed over thirteen hundred hectares and occurred at the same time as the Nakada Marsh reclamation.

Why Nakada Township chose to reclaim the marsh using the township rather than through a coalition of local landlord cooperatives is a complex issue and demonstrates why even Waswo's (1977) and Francks' (1983) analyses of the roles Japanese landlords played in water and land improvement projects over-simplifies the situation. While the former study focused on the role of landlords and the latter on the technological developments they pursued, both studies viewed landlords as the protagonists in development. While it is true that landlords often promoted development projects, these studies minimize tenant usufruct rights over water and land, an essential factor in the analysis. By emphasizing development primarily from the landlord's vantage point, such studies assume social relations based on private ownership of paddies and miss the reality of tenant rights. As a result, the studies end up skewing the dynamic interaction and moral economy between landlords, tenants, and self-cultivators that defined respective relations to the means of production. These relations shall be explored in greater depth in the next chapter.

The major expense was Lancaster pumps, imported through the James Morrison Trading Company in Tokyo. At the time this type of pump was very rare in Japan even in the more advanced agricultural areas. The reclamation itself was also costly, requiring about one thousand wage laborers and many horses, which were needed to form new dikes, canals, and paddies and to expel water from the marsh. Consequently the project provided jobs for the local population, a trend which has carried over into the postwar period and the land improvement projects. The major expenses of the project are listed in Table 3.2.

The Nakada Marsh Reclamation Project transformed Uwanuma into a major rice growing village. After completion of the project, Uwanuma Village gained so much land from the reclaimed marsh that its rice paddy holdings nearly equaled that of Ishinomori.

Table 3.2
Nakada Marsh Land Reclamation Project Major Expenses

Engineering	19,993 yen
Water Sluice	18,405
Bridge	503
Reservoir	511
Land Purchased for Water Canals	2,333
Relocating Houses, Materials, and Wells	285
Office expenses	170
2 Lancaster Water Pumps-@150hp	55,234
Surveying	590
Supervision	5,069
Miscellaneous	2,781
Reclamation	36,684

Source: Nakada Chōshi Henshū Iinkai 1970: 146.

Uwanuma eventually possessed 634 hectares of irrigated land compared to the 714 hectare holdings of Ishinomori.

The ability of landlords in Uwanuma to lease such a large amount of land from the county and then "re-lease" (mata kaeshi) it to tenants, speaks for the entrepreneurial skill of Uwanuma landlords in using the reclamation project to their own advantage. Using their superior position on the irrigation line, these landlords were able to convert collectively held water, marsh, and grass-cutting usufruct rights into individually held property rights over paddy land and labor rights over workers.

Along with the 1921 edict abolishing county control, there arose a problem of how to administer the Nakada Reclamation Project. The Nakada Marsh Cooperative was formed by landlords to provide water for irrigation and to buy the land in order to control it. One of the first projects of this cooperative was to purchase electric motors. In 1927 three 160 horsepower pumps were purchased for a total of 480 horsepower, which was a 180 horsepower improvement. The new pumps were able to pump 147 cubic shaku (1 shaku=0.018 liters) per second, while the old pumps were only able to pump about 100 cubic shaku per second.

STAGE THREE: THE LAND IMPROVEMENT
DISTRICT AS WATER REGULATOR

The Land Reform and the Pump Facility

The Nakada Marsh Cooperative was dissolved in 1949 in accordance with the postwar land reform of 1946. This caused controversy as to ownership of the pump station at Ōizumi, the associated irrigation facilities, and water rights. The land reform did not specifically mention these items because it was directed only at the quantity of land owned and cultivated. At that time the pump station and associated irrigation facilities were worth much more than the land itself and the situation was resolved when the government transferred the pump facilities from Nakada Marsh Cooperative to a new water cooperative called the Ishinomori and Four Other Villages Cooperative (Ishinomori Hoka Yonkason Kumiai). These villages included Ishinomori, Uwanuma, Asamizu, Takarae, and Nagai Village in Iwate Prefecture. The amounts paid during the land reform transfer are listed in Table 3.3.

Table 3.3
Nakada Marsh Cooperative Sale of Assets in the Land Reform

Land	1,377,803 yen 60 sen
Pump	3,360,000 yen 00 sen
Total	4,737,803 yen 60 sen

Source: Kinoshita 1956:61.

The land sale figure in Table 3.3 included the total value of all land in the reclaimed marsh area regardless of whether or not it was cultivated by a landlord or a tenant. The land reform then redistributed the land according to the 1946 Second Land Reform (see Appendix A).

In 1955 Nakada Township was formed according to the new Village Amalgamation (gappei) Law. Ishinomori, Uwanuma, Takarae, and Asamizu became the new Nakada Township. Thereafter, the water cooperative was renamed "Nakada Township Plus Two Other Townships Cooperative" (Nakada-chō Hoka Nichō Kumiai). The two other townships were Hanaizumi Township in Iwate Prefecture and Hasama Township. The gappei administratively separated Mori

Village from Nakada Township. Historically, Mori's roots (at the bottom of the irrigation system) were with Shinden Village, which together with Mori had amalgamated into Takarae Village. With respect to paddy irrigation, Mori stayed with the "Nakada Township Plus Two Other Townships Cooperative" while administratively becoming part of Sanuma, the major city in Hasama Township. The water cooperative later became known as the Five Townships Cooperative (Gokason Kumiai).

Besides it's effect on water allocation, the land reform had great impact on land tenure in the reclaimed marsh area of Nakada Township. During the reform, landlords were forced to sell land in excess of three hectares (combined leased and cultivated land). Most landlords selected the reclaimed marsh paddies to sell. These paddies were chosen because the reclaimed marsh land produced a lower yield and quality of rice than other environmental zones. Because most landlords owned paddies in various environmental zones and because the land reform restricted only quantity of land and not quality, landlords were able to make this choice. This will be discussed in detail in a later chapter.

Immediately after the land reform, water allocation and distribution was very equitable. The Five Townships Cooperative serviced a total of 2,664 hectares of irrigated land as illustrated in Table 3.4. The price of water after the reform remained low. In 1949 the total water tax was sixty yen per tan (0.1 hectare).

In 1951 the farmers formed cooperatives with the objective of constructing ten are rectangular paddies from those which were irregularly shaped. In response to the rising price of rice after the war, many farmers wished to convert field crop areas into rice paddies but in areas such as Asamizu and Takarae where irrigation was traditionally poor, more land was still planted to field crops than to rice.

Table 3.4
Land Service Area in the Five Townships Cooperative

Ishinomori Township	703 hectares
Uwanuma Village	749
Takarae Village	802
Asamizu Village	326
Nagai Village	84

Source: Kinoshita 1956: 67.

In 1949 the government passed a Land Improvement Law making possible the combination of irrigation administration and land improvement. The new administrative units were called Land Improvement Districts, (Tochi Kairyō Ku). The water cooperative soon became responsible for providing water and administrating land improvement projects. While technically under the mayor, it became a semi-autonomous unit that evolved out of the old water cooperatives.

From the early 1960s to the early 1970s the Land Improvement District administered projects to improve drainage facilities and convert dry-land fields to wet-land rice paddies. For Asamizu, however, more wet-land paddies meant that the water shortage problem was exacerbated so a new pump source had to be found. In 1964 a nine meter wide by five meters deep concrete irrigation main was constructed that enabled more water to be pumped from the Ōizumi Pump Station, but it only slightly alleviated Asamizu's water problem.

STAGE FOUR: THE LAND IMPROVEMENT PROJECT

As a result of the Land Improvement Project in Nakada Township, the pump station for Asamizu finally materialized, rice paddy size increased from ten to thirty ares, irrigation and drainage facilities were improved, and fragmented land holdings were consolidated. The northern section of the township opposed the project so Asamizu and Takarae in the south, under the direction of the Land Improvement Office, independently persuaded the national government to sponsor the purchase of a major pump station adjacent to Asamizu. The pump was rated at three hundred horsepower (in comparison to the three 150 horsepower pumps at the upstream Ōizumi Pump Station).

Advancement of the project also gave the inhabitants an improved paddy irrigation system where each paddy is irrigated and drained by a system that is virtually maintenance free. This means that for the first time in Nakada Township history, the downstream villages have better irrigation facilities than the upstream villages.

At present, water regulation for each paddy is one of the most time consuming activities in Nakada rice production and completion of the project will further save labor and promote off-farm employment opportunities. The downstream areas equipped with enlarged thirty are paddies, consolidated plots, and improved irrigation and drainage facilities are about seven years ahead of the upstream areas with respect to the Land Improvement Project. This has given the Asamizu and Takarae areas the opportunity to be the township leaders in utilizing other government programs for crop diversion and entrustment of land which are designed to promote large-scale production.

NOTES

1. In Japanese this poem is:

        Mizu
Mizukara katsudō shite ta o ugokashimuru wa
     mizu nari
Tsune ni onorere no shinro motomete yamazaru
     wa mizu nari
Shōgai ni oite hageshiku seiryoku o baika
     suru wa mizu nari
Mizukara kiyokushite to no odaku o arai
Seidaku aseiruru ryō aru wa mizu nari

By Sōdōshū Kanchō Ryūsensho

2. See Kelly (1982a, 1982b) and Kitamura 1950 (1950, 1973) for excellent descriptions of cultural practices regarding Japanese irrigation.

3. June and July are the months with the highest probability of floods, representing over half the floods since 1600.

4. The years of great famine were 1615, 1624, 1636, 1637, 1641, 1642, 1655, 1669, 1674, 1675, 1696, 1699, 1702, 1704, 1707, 1713, 1714, 1720, 1729, 1732, 1734, 1744, 1757, 1759, 1763, 1765, 1780, 1783, 1785, 1788, 1833, 1834, 1835, 1836, 1837, 1839, 1841, 1842, 1843, and 1869 (Tomegunshi Henshū Iinkai 1924).

4

Land Rights

INTRODUCTION

Changes in land and water rights paralleled the development of the irrigation system described in the last chapter. These rights evolved from use rights into rights of private ownership at the village, hamlet, and kin level of social organization. They have continually rested on the moral values of each community (Brow 1981, Firth 1951:141; Sahlins 1972:198; Scott 1976: 2-6), and are inextricably interwoven with rights over human labor (Polanyi 1957:178).

By 1983 private property rights had become more predominant, but usufruct rights over land and labor are still present to a minor degree. Land rights, both use rights and private property rights, reflect past, present, and future social relationships regarding land and labor. A full explanation of these rights entails more than just rental rates or marginal utility of action undertaken by landlords or tenants. It must include the totality of responsibilities and obligations of each party towards each other and the land.

The political and economic significance of tenant rights is intertwined, as seen in the critiques of Richard Smethurst (1986) by Herbert P. Bix (1987) and Masanori Nakamura (1988). To this we must add the possibility of kinship ties. In Nakada Township it was common for tenant rights to be bound up in the kinship affiliation because main and branch households were often in a landlord and tenant relationship. Both foreign and native scholars have failed to adequately describe tenant use rights (kosakuken) as opposed to land tillage rights (kosakuken) and this

51

has hindered our understanding of the social realities of rural Japan. The fact that farmers in the 1980s were conscious of past tenant and tillage rights plays a large part in understanding the reluctance of farmers to lease their land to others despite government financial incentives to do so. This is especially true in comprehending the essence of the land reform or the farmers aversion towards participating in the various lease projects promoted by the government to increase the scale of farming such as the Land Improvement Project described in Chapter 10.

Not only do English ethnographies on rural Japan omit this aspect of rural social relations but so do the works which deal with the 1920s when the tenant right was consolidated. For example, Waswo (1977), Francks (1983), and Smethurst (1986) all concentrated on prewar Japanese agricultural development, and were concerned with the number of tenants, the amount of rent, and number of tenancy disputes rather than the land rights themselves, which were often at the core of the debate. Although the idea of tenant rights was mentioned in Dore's <u>Land Reform in Japan</u> (1959:42, 64) and dismissed as affecting less than one percent of the total tenanted land, the concept was probably not developed because the American land reform in Japan was directed at abolishing absentee landlordism, regulating the amount of land tilled, and promoting democracy among small-scale owner farmers. In regard to the prewar tenancy system in Japan, Dore does, however, explain paternalistic landlords in detail. It seems probable that the planners of the American occupation were not aware of the tenant right. In Chira's (1982) work on the architects of the land reform, the only mention of the tenant right is in an appendix taken from Ogura (1982).

In 1989 the <u>kosakuken</u> applied to between one and two percent of all paddy land as a result of postwar laws that have redefined the landlord-tenant social relationship. Nevertheless, on paddies where the tenant right predates the newer legislation, the monetary value of the tenant right is presently equal to about one-half the value of the paddy itself. It should be remembered that tenants also had a right to inherit the tenant right and pass it on to the next generation. Compensation (<u>risakuhōshōkin</u>) to the tenant was necessary in the event that the landlord terminated the arrangement. In 1983 when a landlord-tenant relationship was discontinued on paddies to which the tenant right applied, instead of cash compensation the tenant sometimes was able to split the land and obtain full private property rights on his half in exchange for relinquishing his inheritable right to till the land on the other half.

Past and present usufruct rights affect present rental rates. For instance, rental rates for all paddies are a function of the type of land right agreement between tenant and landlord: the stronger the land right, the less the rental rate. Table 4.1 explains the relationship between the period when the rental agreement was initiated and the actual rate amount, and whether it was paid in kind or in cash. The lowest rental rates were for land which had been tenanted between the early 1920s when the

Table 4.1
Type of Rental Payment and the Date of Initial Rental Agreement

	In Cash Only	In Kind Only	Cash & Kind	Other
Before the land reform	93.1%	5.2%	0.6%	1.1%
After the land reform but before 1960	79.4	13.4	0.5	6.8
Between 1960 and 1970	47.8	48.6	0.1	3.5
After 1970	59.3	35.8	0.7	4.2
Average	70.2	25.8	0.6	3.5

Source: Zenkoku Nōgyō Kaigisho 1980.

tenant right was strengthened due to peasant protests, and the abolishment of the tenant right in 1970 with the Amendment to the Agricultural Land Act of 1952 (see Appendix A).

In the event that the rent was to be paid in kind, it constituted a fixed amount of the crop calculated in sixty kilogram bags. Since the threshing was done in the compound of the tenant, however, there was no way for a landlord to actually determine if the rice came from his plot. Therefore, within certain limits of taste and grade, it was up to the discretion of the tenant to give the landlord a high or low quality of rice.

Rent for land tenanted before the land reform was usually paid in cash, as shown in Table 4.1. This demonstrates the strength of the tenant right because by paying cash the tenant was then eligible to receive all government subsidies on the marketed rice. Since 1960, however, there has been a trend rice to Nōkyō under their own name and received the government subsidies. Table 4.2 reveals the effect of the type of tenant right on the rental rate regardless of whether it was in cash or in kind.

Paying rent in kind also facilitated the practice of unregistered tenancy in which a landlord was registered with the Land Commission as if he were actually tilling the land. In reality the landlord would verbally contract with a tenant to farm the land.

Table 4.2

Land Rights and 1981 Paddy Rent Rates: 1981 Average Paddy Rents According to Time of Initial Rental Agreement

Date of Initial Rental Agreement	Rental Rate (Yen)
Before the land reform	10,796
After the land reform but before 1960	13,765
Between 1960 and 1970	30,349
After 1971	34,800

Source: Zenkoku Nōgyō Kaigisho 1980, cited in Kondō 1981:382.

The Nakada data correlate with the above national trend. Claiming rising real estate taxes, landlords have gained considerable power controlling agricultural rental rates through informal (non-registered) rental agreements with tenants. After 1960, most landlords tried to compensate former tenants in order to abolish tenant rights. In many cases it meant sacrificing a value equal to one-half the land, which went to the tenants as reparation. New short-term lease agreements were applied to the remaining land. It was on this land, where unregistered tenancy was common, that landlords applied the highest rents, that were to be paid in kind. The rental rate for this type of agreement varied between four and five bags of rice on land that produced an average of nine to ten bags per year.

Because rents could be paid in various combinations of cash or in kind, further investigation into the background surrounding usufruct land rights in Nakada Township during the Tokugawa

Period is merited. The remainder of this chapter will explore the development of tenant usufruct rights, explain the types of land rights held in 1800 by Cherry Tree Village in which K Hamlet was located, and give examples of how the tenant usufruct right has evolved into a private property right during the twentieth century.

THE DEVELOPMENT OF USUFRUCT RIGHTS

The idea of preserving and maintaining land actually predates the Tokugawa Period and was predicated on the ideology that the national land of Japan was to be preserved by the people for the emperor. During the Tokugawa Period, usufruct land rights developed when domain land was given to retainers. At that time the ownership rights to the land were clearly in the hands of the retainers, while the actual cultivation was done by the peasants. Peasants had considerable autonomy to farm the land as long as they paid the rice tax, based on the amount and productivity of the land they cultivated. The dynamics and bargaining power of this relationship is revealed in the peasants petition later in this chapter.

In 1645, with the exception of castle town land, the Tokugawa government prohibited the buying and selling of land. In 1711 it unsuccessfully forbade the division of land upon inheritance (Ōno 1977:85). The main reason for the prohibition was to prevent retainers and lords from consolidating holdings and posing a threat to the feudal order. The ideological basis for the edicts was the concept of preserving and maintaining the land (tochihoyū) for the emperor even though the feudal vassals had gained private rights to the taxation of domain land. The peasants tilled both retainer and non-retainer land. Edicts prohibiting buying, selling, and fragmenting of land through inheritance seem to have been more relaxed in the countryside where land and labor were exchanged more often by oral agreement than by formal written contracts. In fact, by the end of the period the peasant-domain relationship was seriously strained (see Appendix A).

Although Dore (1959:64) dismisses permanent tenancy (eikosaku) as affecting less than one percent of all tenanted land, it appears that there was a great amount of regional and local variation with respect to usufruct rights over land. Ōno (1976:124) and Nōchi Kaikaku Kiroku Iinkai (1951:232) refer to many forms of usufruct, each with local names, and describe great diversity of local customs. This would be consistent with the variability of traditional labor forms as described by Thomas Smith (1959). Some of the most common land rights were secured through peasant involvement in land reclamation and improvement projects and it appears that the domain governments bent the rules to authorize the granting of usufruct rights, particularly when they sought to increase their tax base by sponsoring such projects. It is likely that this was the case for Cherry Tree Village.

Generally, permanent tenant rights to till the land included: (1) the permanency and inheritability of the arrangement; (2) the right of the tenant to sublease the land; (3) the right of the tenant to end the arrangement by choice or by not paying rent; (4) the right to a lower rent rate than prevailing short term tenancy rates; and (5) in some cases the responsibility to pay tax. According to Ōno, restrictions on buying and selling traditional rice paddies, honden (also called honchi in the survey to follow), were more severe than on newly reclaimed lands, shinden. According to the Cherry Tree Cadastral Survey (shūmon aratamechō) of 1800, approximately equal amounts of honden and shinden were maintained in neighboring villages by Cherry Tree Village. Apparently, the restrictions either were not enforced or were bypassed. This was largely due to the fact that Cherry Tree Village enjoyed relative control over irrigation water.

A matter further complicating the analysis is that many communities in Tōhoku have a kinship structure based on main honke and branch bunke households, with the latter working the land of the former at the time of establishment of the branch household. In many cases, these close household ties have been maintained for generations. It must be imagined that then as now complex economic and social exchanges cemented these ties and that many types of usufruct rights were continued through the kinship mode.

Although the records of rental rates in other villages are incomplete, it is clear that in 1800 members of Cherry Tree Village were paying tax on land holdings in other villages. Nearly one-quarter of the land tax paid by Cherry Tree Villagers was based on extra-village land holdings. The total tax on traditional rice paddies (honchi) held in other villages was 19 kan and 95 momme (1 kan= 3.75 kilograms and 1 momme=3.75 grams). This was about equal to that of the total newly reclaimed lands (shinden) which was 18 kan and 901 momme. All the villages listed except Uwanuma were downstream on the irrigation system, indicating that these villages, by 1800, were beginning to lose their land rights due to lack of control over irrigation water.

USUFRUCT MARSH RIGHTS BY VILLAGES

The usufruct land rights to the marsh and its resources are of critical importance to understanding the changing relationships between the villages. Evidence points to the probability that marsh resources were used as a commons by villages and that these rights varied according to the respective village power relationships. Usufruct rights to that which is above the water are mentioned in one contract made in 1897 and no doubt distinctions were made concerning the rights to resources above and below the water surface, at the edge of the marsh (marsh grass) and in the marsh proper, and to types of water (e.g. stagnant versus fresh, warm versus cool) flowing from the marsh for irrigation use.

According to an official 1720 map, the marsh was divided between Uwanuma, Ishinomori, and Nishi Nagai Villages. Uwanuma had rights to about three-fourths of the marsh, including most of the area designated as yachi, which was undeveloped marshland from which grasses for horse fodder and composting were cut. Ishinomori and Nishi Nagai each had rights to about equal portions of the remainder. Cherry Tree Village did not have rights to the marsh but did have sizable rice land holdings and a small amount of yachi that extended from the marsh area into the village's own rice lands.

By the end of the nineteenth century village usufruct rights to land and water were being converted into legal rights. During this process, a new relationship over the use of marsh resources developed. First, the marsh was classified as publicly owned land (kanyuchi) by the new Meiji government, putting it under the jurisdiction of the county (gun). Second, the village of Cherry Tree (just before its 1885 amalgamation into Uwanuma Village) jointly shared the southern part of the marsh with Ishinomori Village. During the 165 year interim between 1720 and 1885, Cherry Tree Village obtained additional rights to the marsh. The amount of land in the marsh allocated to Uwanuma decreased to less than that held by Ishinomori and Cherry Tree Villages combined. According to a legal document signed in 1907,

Ishinomori Village held ninety-nine hectares of the marsh at the time it sold its kanyūchi rights in order to acquire public domain (kōyū) irrigation water. Nishi Nagai Village's area remained the same. Since the marsh was to be administered by the county, the local village governments were responsible for collecting fees from grass-cutters. Regulation was necessary for this precious commodity which was used both for horses and composting as green manure. In 1904 there were 110 registered grass cutters in Uwanuma Village, many of them coming from a hamlet which lies at the edge of the yachi and is directly upstream on the irrigation canal from K Hamlet.

During the last part of the nineteenth century and first decade of the twentieth century, a great debate ensued between Ishinomori and Uwanuma (into which Cherry Tree Village amalgamated in 1885) concerning use rights over grasslands. No doubt the key issues in the debate were access to irrigation water and the right to cut grass. First, Uwanuma controlled the irrigation source at the Oizumi Intake Sluice on the Kitakami River, and consequently regulated water flow and grass growth. Second, Uwanuma Village controlled the flow of water into the marsh. By raising or lowering the water level in the yachi, Uwanuma could control the yachi growth and harvest periods. Third, Uwanuma villagers possessed more horses for which cut grass was necessary. According to the Nakada Marsh Papers[1], income from Uwanuma Village yachi grass cutter fees was approximately three times that from Ishinomori.

THE DILEMMA OF THE LOCAL LANDLORDS

As Kinoshita (1956:23) points out, local landlords were losing power in the years prior to the reclamation of Nakada Marsh. Table 4.3 depicts an increase in Nakada land ownership by landlords from outside the township and shows a corresponding decrease in local village control during the seven years preceding the reclamation of Nakada Marsh. In Takarae Village, downstream on the irrigation line, there was a decrease of 123.7 chō while in upstream Uwanuma there was a much smaller decrease amounting to 33.1 chō. The sharp difference is in part due to Takarae's vulnerability to frequent and severe flooding such as the flood in 1876.

Extra-township control became a threat to local landlord control and increased the bargaining power of the tenants vis-a-vis their local landlords. This is probably one of the reasons why landlords initiated the reclamation project through county auspices. As stated in the preceding chapter, most reclamation and land improvement projects were organized by and for landlord cooperatives. The Nakada example, however, was structured to benefit the county directly and the landlords only indirectly. In other words, gains by local landlords were less than usual.

Table 4.3
Early Meiji Ownership of Land in Downstream and Upstream Villages on the Irrigation Line (in chō; 1 chō= 0.997 hectare)

Land Ownership Category	Takarae 1900	1907	Uwanuma 1900	1907
Same Village	904.2	780.5	955.1	922.1
Other Village, Same County	473.8	518.2	126.7	171.3
Other Village, Other County, Same Prefecture	19.0	18.9	2.7	3.8
Other Village, Other County, Other Prefecture	5.0	62.0	19.7	20.0

Source: Kinoshita 1956:33.

THE TENANT'S RIGHT AND THE RECLAIMED MARSH

Following the 1908 Nakada Marsh reclamation, peasants were able to negotiate for cultivating rights from the local landlords. At this time actual tillers had to struggle thigh deep in mud in order to transplant the rice. Hence landlords had much to gain by compensating these tenants. In 1923, tenant rights over the marsh area were redefined, with tenants gaining back some rights lost since the Meiji Restoration of 1868. A tiller's fee (okoshi chin) of about three yen was placed on each ten ares where the yield was equal to a cash equivalent of fifty to sixty yen. Accordingly, the rental rate was less than five percent of the yield, a significant decrease from the 1921 official rate which was nearly thirty-six percent of the harvest. (In 1921 the rent, to be paid in cash to the county, was the cash equivalent of 6.1 to on paddies which averaged a yield of 16.81 to.) Because the practice of subleasing was common, the actual rental rate was slightly higher because each person usually made profit on the practice of leasing it to someone else. In effect the "tillers" who subleased out their parcels were actually landlords. Nevertheless, they maintained their rights as tenants because they were registered as such and therefore continued to pay the okoshi chin to the county cooperative. These tenant rights were inheritable and could be bought or sold by the "tenant."

By 1937 the tenant's right was valued at approximately forty to fifty yen per ten ares and ranged from twenty yen to one hundred yen, depending on paddy location (Nomura 1937: 72). In addition, land cultivators were allowed to keep 2.3 to (10 go= 1 shō; 10 shō= 1 to; 1 to= 0.1 koku; 0.1 koku= 15 kilograms) of the average yield (Nomura 1937:173).

During the land reform, ownership of all land in the reclaimed marsh area was returned to the actual tillers. At that time permanent tenant rights to the reclaimed Nakada paddies disappeared. The land reform did not effect all areas to which permanent tenant rights applied as there were still areas within the other environmental zones in which these rights existed.

THE POSTWAR CHANGE OF THE TENANT RIGHT INTO PRIVATE PROPERTY RIGHTS

As noted in the beginning of this chapter, the average national land rental rate to which permanent tenant rights applied was approximately one-third the going rate. This also applied to Nakada Township. Article 20 of the Agricultural Land Law (Nōchihō) required compensation (risakuhoshō) be given to tenants who relinquished their tenant right.

In order to charge higher rent rates, landlords in K Hamlet were required to compensate tenants in thirteen cases from 1958 to 1983. According to the Land Commission registers, reimbursement was given in the following cases: forced selling through eminent domain when the government needed to improve roads or to construct a bridge, responsibilities to relatives, retirement of either the landlord or tenant, and the inability of the tenant household to till all the land due to (non-farm) job responsibilities.

In five of the cases, the tenants received land ownership rights equal to approximately one-half of the tenanted acreage as their compensation. In the remaining cases, the tenants received cash reparations equal to about one-half the value of the land.

Most cases listed by the township did not involve immediate branch or main household relativess However, in such cases where there were close household alliances or relatives, the amount of reimbursement was often less than in other cases.

In contrast to reasons stated on paper, interviews revealed that, almost universally, landlords, pressured by rising real estate taxes, were asking their tenants to relinquish tenant rights so that landlords could charge new tenants higher rents on the land. The new tenant-landlord relation was legitimized through the 1970 Amendment to the Agricultural Land Act of 1952 and favored the landlords by abolishing the inheritable tenant right. Because of the existing social arrangement based on permanent tenancy, it would have been morally unethical for these landlords to ask for higher rates to tenants that had the tenant right. This was even truer for close relatives who were tenants. A landlord doing so would have risked losing social status.

NOTES

1. Rather than footnote each document separately, I am referring to these as the Nakada Marsh Papers. These volumes of papers and contracts are maintained in register form in the Land Improvement and Land Commission Offices. Other papers and contracts were privately held and shown to me when I interviewed farm households.

5

Land Fragmentation and Consolidation

> Once there was a farmer who sat in his fields while counting his scattered paddies. He came up one short in his calculations and was perplexed until he realized that it was the one he was sitting on. (Told to me by Mayor Haga of Nakada Township.)

LAND FRAGMENTATION

Land holdings in Nakada Township and the Tōhoku Region are fragmented, with each household owning an average of 1.1 hectares scattered in seven or eight plots. In recent years this has been a major stumbling block towards achieving a more "rational" agriculture (reducing costs per unit produced). The major causes of land fragmentation included: (1) household fissioning in separate environmental zones; (2) a shift from single heir to multiple heir inheritance; (3) increases in the rate of land transfer; and (4) the 1946 Land Reform.

Household Fissioning

The single greatest cause of land fragmentation was the formation of new branch households, each receiving land from the main household. Rice and vegetable crops were raised in different environmental zones, so it was necessary for the main household to allocate to its branch household land in more than one environmental zone. Thus, the holdings of the main household were gradually chipped away leaving only one or two plots in any one area. Figure 2.4 in Chapter 2 demonstrates the break-up of a large holding in the traditional rice growing zone by a main household for branch members of the edōshi group. This edōshi group was comprised of one main household and ten branch households. Figure 5.1 shows the 1983 land holdings of the same edōshi group's main household. Eight non-contiguous small plots are shown in three environmental zones. Four of these remain in the traditional rice growing zone.

Environmental zones play a significant role in household fissioning with only nine of the fifty-six farming households in K Hamlet tilling land in only one environmental zone. Both rice and field crops such as wheat, oats, and soybeans were needed so prewar farm households had to cultivate land in more than one environmental zone, enabling them to take advantage of varied soil types, climatic conditions (i.e. sunlight), and irrigation and drainage facilities. For growing rice, differential water availability according to ecological zones was adaptive, since household labor could be more fully utilized. This was true particularly during the transplanting and harvesting seasons when demand for labor peaked. These environmental zone combinations are illustrated in Table 5.1.

Table 5.1
Environmental Zones and the Land Holding Pattern

Individual Household Land Holding Pattern	Number of Households
Reclaimed Marsh Zone Only (Numachi)	4
Traditional Rice Zone Only (Kiseiden)	1
Traditional Field Crop Zone Only (Hatake/Kaiden)	4
Reclaimed Marsh Zone and Traditional Rice Zone Only	4
Reclaimed Marsh Zone and Traditional Field Crop Zone	2
Traditional Field Crop Zone and Traditional Rice Zone	19
All Three Zones	22

Note: The above comparison is made without considering the households or the environmental zone next to the river.

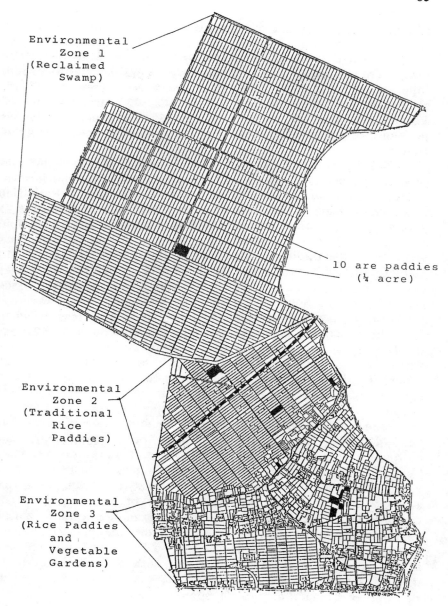

Environmental Zone 1 (Reclaimed Swamp)

10 are paddies (¼ acre)

Environmental Zone 2 (Traditional Rice Paddies)

Environmental Zone 3 (Rice Paddies and Vegetable Gardens)

Source: Moore 1985:136.

Figure 5.1
Fragmented Land Holdings of an Edōshi Main Household

Mode of Inheritance and Succession

A second cause of land fragmentation has been the trend away from single heir inheritance toward multiple heir inheritance. At the same time there has been a gradual separation of the concepts of succession and inheritance. Formerly it was the right of a person who succeeded to the position of household head to also inherit the household resources.

The Meiji Civil Code of 1898 provided that one-half the physical property of the household must accompany succession. Nakada land transfers at the turn of the century show that in many cases death of the former head, succession to headship by a new heir, and inheritance of the land occurred within a year of each other. This demonstrates the key role played by land as a resource in the traditional structure of the household. The Meiji Civil Code also required that land be held by males. This gave considerable power to "adopted husbands" (mukoyoshi) who married into the household). In some cases, however, the actual transfers skipped a generation (tobikoshi sozoku) in avoidance of the inheritance tax or to preclude giving land to a mukoyoshi. In the case of tobikoshi sozoku the property relations associated with individual household positions such as household head were subordinated to the economic interests of the group. This is further evidence indicating the flexibility of the household system.

The Civil Code of 1948 abolished the position of household head (konushi) in favor of family head (setainushi) and encouraged divided inheritance. Early postwar succession taxes favored divided inheritance by assessing single succession at a higher rate. Nevertheless, the new law did leave loopholes that enabled the household to maintain most of its property intact through the inheritance process. For example, a family head could exclude one-half of the estate from the prescribed legal distribution, which held that a male spouse received one-half and female spouse one-third of the estate with the remainder to be divided among the children. In 1981, the law was changed so that either a male or female spouse received one-half of the estate.

The rights of "adopted husbands" were diminished in the postwar Civil Code by giving yoshi (including mukoyoshi) the same rights as children, which is to say that de jure rights for adopted husbands decreased from full ownership of the land to only partial inheritance. In fact, after the land reform there was a marked increase in female heirs registering land in their own name. This is particularly true in the case of a "marrying-in husband" mukoirikon.[1]

At first it would appear that equal inheritance and succession laws should have caused considerable fragmentation of land holdings. This has been circumvented through two legal loopholes. While the Meiji Civil Code of 1898 focused on inheritance and succession occurring at the same time and usually upon death, the postwar Civil Code provided opportunities for succession and inheritance to be treated separately before death. The first method is a written last will and testament (yuigon) enabling the bequeathment of property to a designated heir or dividing it between designated persons. The second method seeks to elicit the cooperation of the other siblings. In this case it is possible for either a "conciliatory agreement on the division of inheritance" (yuisan bunkatsu kyogi) or an "abandonment of inheritance" (sozoku hoki). In either case, the family makes an agreement on how the inheritance should be divided. Agreements can take on a legalistic tone with personal seals affixed but are often simply verbal agreements. Usually non-successors receive money, a lifetime supply of rice if they move to the city, elaborate weddings, or a combination of all three. As shown in Photo 1.1, rice plays a prominent role in the wedding of the successor to a farm in K Hamlet. Rice is placed in a container to symbolize the continuity of farm holdings.

The traditional mode of inheritance, transferring land without fragmentation upon the death of the household head (sozoku), has rapidly decreased in importance since the early 1970s. At that time sozoku inheritance represented about three-quarters of all inheritance cases. Sozoku inheritance had the advantage of liberal inheritance exclusions enabling the average farm to be inherited with only minor taxation problems. In the mid-1970s two things happened which decreased the incidence of sozoku inheritance to about one-quarter of all cases.

First, land values soared to a point beyond the value of inheritance exclusions. The value of paddy land in Miyagi Prefecture is shown in Figure 5.2. Farmland value and the capital assets of the household (especially savings) caused the total estate valuation to exceed the inheritance deduction, which had to be raised in 1964, 1968, 1975, and 1988. In 1988 a typical Nakada farm household's land holdings of 1.2 hectares (three acres) was worth approximately twenty-four million yen

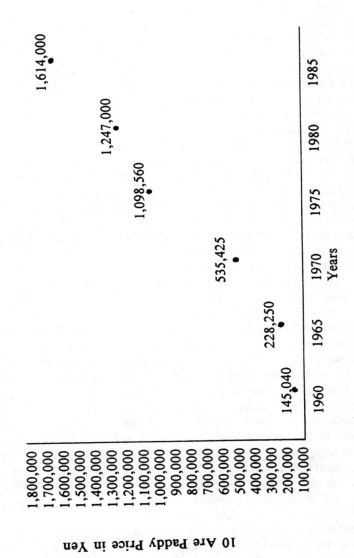

Source: Miyagi-ken Nōgyō Kaigi 1983a.

Figure 5.2
The Value of Paddy Land in Miyagi Prefecture

($192,000). Prior to the inheritance law's 1988 amendment, the exclusion was twenty million yen raising the value of the farmhouse plus any bank savings over the 20 percent inheritance tax bracket. With the 1988 amendment to the sōzoku inheritance law, there will be an exclusion of forty million yen. Farming households will then be more likely to opt for sōzoku inheritance again.

Because of rising real estate prices, it was to the advantage of the household to transfer the land as soon as possible to take advantage of the lower value upon which the tax was assessed. This was a clear departure from the prevailing idea in the Meiji Period to postpone the tax as long as possible.

The second reason for decreased use of sōzoku inheritance was a 1975 revision in the gift tax law which provided a "tax grace" (nōzei yūyo) to living individuals who gave "lump-sum" undivided farmland gifts to an heir over the age of eighteen and possessing three years of farming experience. These lump-sum gifts (ikkatsu zōyo), aimed at preserving family farms intact without fragmentation, qualified for a deferred tax payment for twenty years or until the next occurrence of inheritance (whichever came first).[2] The trend away from sōzoku inheritance towards zōyo gift transfers is recorded by the Land Commission in Nakada Township which is required to list reasons for land transfer. In 1953, for example, the number of zōyo paddy transfers was about equal to the number of all other paddy transfers. In 1983, twenty years later, however, zōyo transfers were about three times as common as all other categories of land transfer combined.

This has had a mixed effect on land fragmentation. While the goal of the lump-sum gift tax revision was to maintain farmland intact, the change in the gift tax generally accelerated land transfer. The sōzoku inheritance tends to slow down land transfer because it occurs at death but "gift" (zōyo) inheritance takes place while one is alive and hence provides for more transfers over time. However, many farmers feeling tight financial pressures have been unable to take advantage of the lump-sum tax deferments. These farmers have had to sell pieces of their land and have mixed sōzoku and zōyo inheritance methods.[3]

Recent government policies, such as the agricultural retirement system of 1970 (see Appendix A) have favored lump-sum gift inheritance. The agricultural retirement system requires that the land be transferred before the recipient can receive retirement benefits. When asked how and when they planned to transfer their land to their successors, farmers responded "agricultural retirement system," which meant at age sixty-five in a lump-sum to a designated heir. Lump-sum gift inheritance comprised between 80 and 90 percent of all gift inheritances (zōyo) in Nakada Township in 1980. Ten years earlier lump-sum inheritance represented only about one-half of the gift inheritances, with partial inheritance making up the remainder. Eighty percent of lump-sum inheritance occurred because farmers are using the agricultural retirement system. This system places an enormous burden on the household by forcing inheritance to occur within the specified twenty year time limit of the program and in that sense makes the ie less flexible than under the Meiji Civil Code.

As noted in Chapter 8, the "heir problem" (kōkeisha mondai) has become a national problem in rural Japan and is usually attributed to out-migration from the rural areas. It is also partly due to the need to find an heir and transfer the land within the twenty year limit required by the agricultural retirement system. Traditionally the ie could wait until the death of the household head before transferring the land, and so had the time to shift personnel, reduce or increase its labor resources, and try out potential heirs. The government's efforts to "rationalize" the land holding pattern by speeding up the rate of land transfer is a direct attack on the resource base of the ie and it is understandable that part-time farmers, wishing to maintain maximum flexibility, have been reluctant to join the program.

Land Transfer Rates

Government policies have been successful in increasing both the rate and amount of land transfer. Since both the rate and amount have been so low, even greater policy changes are necessary to effect the kind of transfer rates that government planners desire in order to create large-scale farming. Figure 5.3 shows a gradual increase in the rate of general land sales in Nakada Township and number of cases involving financial difficulties.[4] According to Land Commission records, a total of 9.1 hectares was sold in 1985. Because there are a total of 4,090 hectares of agricultural land in Nakada Township, a sale of 9.1 hectares per year would mean that it would take 449 years for

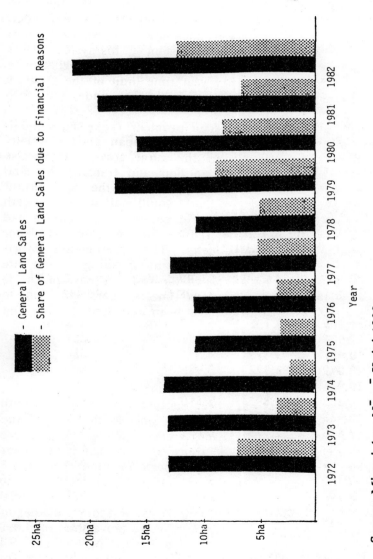

Source: Miyagi-ken Nōgyō Kaigi 1983a.

Figure 5.3
General Land Sales and the Number of Financial Difficulties

the whole township to be sold to new households. The numbers for Miyagi Prefecture are similar. Table 5.2 lists the number of cases and acreage of private land ownership transfers for the prefecture. There were 152,800 hectares of arable land in cultivation in the prefecture in 1985 but only 675 hectares of land changed hands (Nōgyō Kaigisho 1986:50), a mere 0.4 percent of the total. Also in 1985, nationally 0.4 percent of the land was sold. Land transfers for Miyagi Prefecture and the nation are given in Table 5.3.

Due to industrial development, some land prices in Nakada Township have risen dramatically, making agricultural land a good investment well above the inflation rate. Households desiring to depart from farming, and particularly those who have had financial difficulties, account for the majority of the land sales.[4] Land sales due to financial problems are particularly prone to cause land holding fragmentation because small rather than large pieces are sold off and the farmers who want to buy the pieces often do not own contiguous pieces.

Table 5.2
Farm Land Ownership Transfers
in Miyagi Prefecture (in hectares)

Year	Number of Transfers	Acreage	Compensated Transfers	Compensated Acreage
1972	11,109	3,195	5,517	873
1973	12,901	3,973	6,281	1,281
1974	10,951	3,313	4,504	684
1975	10,079	3,322	3,985	560
1976	10,986	4,217	4,223	732
1977	10,285	3,895	3,860	556
1978	10,088	3,636	3,463	501
1979	9,876	3,575	3,452	545
1980	10,543	3,941	3,805	677
1981	10,941	3,642	3,995	634
1982	10,745	3,700	4,018	761
1983	9,304	3,306	3,444	536
1984	9,017	3,386	3,353	657
1985	8,747	3,217	3,043	512

Source: Miyagi-ken Nōgyō Kaigisho 1985:15.

Table 5.3
1985 Farm Land Transfers in Miyagi Prefecture and Japan (in hectares)

Type of Transfer	Miyagi	Japan
Compensated transfer of ownership title	675	24,305
Non-compensated transfer of ownership title	2,478	50,936
Transfer of tenant right	55	1,374

Source: Nihon Nōgyō Nenkan Kankōkai 1988:530.

Note: All the above categories were transferred in accordance with Article 3 of the Agricultural Land Law.

The Land Reform

The land reform was a major factor in land fragmentation, second only to the branching process. The reason the land reform played an important role was that it did not make a distinction between quantity and quality of land owned. As a result, when landlords were required to limit the amount of their tenanted land to one hectare, they usually chose to abandon land that was non-contiguous to their other plots and located in the least desirable environmental zones, a situation which will be discussed in Chapter 6. This occurred particularly in areas of local landlord control, such as in K Hamlet. In K Hamlet, the reclaimed marsh paddies were abandoned, leaving the landlords with the land closer to the hamlet, which was more highly valued. Although there was a well-developed land valuation system in prewar Japan that took quality of land into account, the quality of land was not considered by the American Occupation reformers. As a result, areas with the highest degree of absentee landlordism typically had higher rates of land fragmentation after the reform.

When we compare Uwanuma (which had a high degree of prewar resident landlord control) and Takarae (which had a high degree of prewar absentee landlordism) we see the fragmenting effect of selling land for financial reasons that must have occurred more often in Takarae. Demonstrating a lower degree of fragmentation, the fifteen Uwanuma hamlets cultivated land in fewer land districts than did the Takarae hamlets. Fifty-four percent of the Uwanuma households and 37 percent of the

Takarae households cultivated land in four or fewer land districts. (Miyagi-ken Tome-gun Nakada-chō Nōgyōiinkai 1980).

Upstream-Downstream Parcel Subdivisions

Land in Nakada Township is historically more fragmented within downstream districts due to the tendency to subdivide plots in areas that had insufficient irrigation water. The 1889 land maps (kiriezu) of Nakada Township reveal that subdividing was advanced in areas lacking a stable supply of irrigation water. It is often assumed by farmers that subdividing enhances labor intensive agricultural conditions that result in a higher production output. Water levels could be held more constant in subdivided paddies and the laborers could take advantage of ecological variations in soil, wind, and sun conditions. It is probable that in times of water shortage, some of the paddy subdivisions could be sacrificed while others are maintained. Also, downstream areas, being less able to afford irrigation improvement projects, had more to gain from subdividing.

In Table 5.4, subplots per paddy in 1899 for a well-irrigated upstream paddy district in Uwanuma are compared[5] with two districts at the bottom of the irrigation line (one district had the distinguished name of "Buttocks of the Irrigation Line"). The paddy district with the reliable irrigation supply had a much higher area per subdivision ratio than did the others.

Table 5.4
Paddy Subplots and Position on the Irrigation Line in 1899

Land District	Subplot Area (in square meters)
Upstream Irrigated District	355.78
Downstream Poorly Irrigated District 1	194.53
Downstream Poorly Irrigated District 2	266.84

Source: Nakada-chō 1892.

Land Reclamation Projects

Land reclamation projects are also associated with high land fragmentation rates for several reasons. First, when a reclamation project was undertaken it usually required cooperation of people on either the village or township level. When the project was completed everyone within that village had the opportunity to buy land. The Nakada Marsh Reclamation Project, for instance, was a county project and those who bought the tilling rights were mainly from the four surrounding villages.

Land reclaimed through such projects is often lower in value because soil conditions and drainage are often poor and this is often sold first when farmers encounter financial problems. In general, the transfer rate is higher for recently reclaimed areas.

Land Use

Traditionally, a large percentage of production was for subsistence and a smaller amount was for livestock feed. Therefore, it was necessary for farms to use integrated farm management (fukugo keiei) that included production of rice, wheat, oats, soybeans, vegetables, and livestock. To do this, farmers required land in various environmental zones. According to individual household data collected in the 1955 Census of Agriculture for K Hamlet, the majority of households still grew a combination of the above. Much of the crop was not sold and a greater amount was held back than sold in every crop category. Most was probably used directly on the farm, in barter for other goods, and for social distribution to friends and relatives.

In 1955 rice comprised 79.9 percent of all agricultural sales, including livestock and vegetables. In 1983, however, rice accounted for about 90 percent of all agricultural sales. Even households with the lowest percentage of rice sold compared to rice produced marketed over two-thirds of their production. This demonstrates a trend towards monocropping and away from mixed cropping and self-sufficiency.

The social distribution of rice remains important. In 1983 the social distribution of rice to non-heirs and relatives in cities ranged from two to four bags (each bag weighing sixty kilograms). The average farm produced about ninety bags of rice each year so this accounted for about 4 percent of the yield.

Table 5.5
Percentage of Crop Sold by K Hamlet in 1955

Crop:	Percent Sold/ Total Produced:	Sales Amount In Yen:
Rice	48.9%	4,723.91
Wheat	34.0%	296.80
Oats	23.4%	397.50
Soybeans	35.2%	427.55

Source: Nōrinsuisanshō Keizai Kyoku Tōkei Hōkokubu: 1955 (original survey sheets stored at Tōhoku University).

Vegetables were also distributed to neighbors, especially when the yield was "first pickings" (hatsumono).
   Households make a clear distinction between rice that they eat or socially redistribute and rice that they sell. Figure 5.4 reveals plot locations for rice consumed by K Hamlet households. The majority of households chose what they considered to be the best tasting rice for themselves. There was a broad consensus that the middle environmental zone produced the best tasting rice and households which were not lucky enough to cultivate land in that zone were forced to select the best tasting rice from the other zones. The poorest tasting rice came from the zone containing the reclaimed marsh. Production methods also differed in the zones, with natural sun drying producing the best tasting rice. Some people said that they could distinguish between the color of sun dried rice and rice which had been dried in a machine dryer. A further discussion of rice production and taste is found in Chapter 8.

LAND CONSOLIDATION

Projects to Consolidate and Rationalize Land Holdings

   There have been four projects in Nakada Township for the purpose of consolidating and rationalizing land holdings. While all four projects were very successful in rationalizing the shape and use of the paddies they were not effective in consolidating the holdings for either tillers or owners. In chronological order the projects were: (1) The 1933 Project to Change Vegetable Fields to Rice Paddies (Hatake Kaiden Jigyō); (2) The 1951 Land

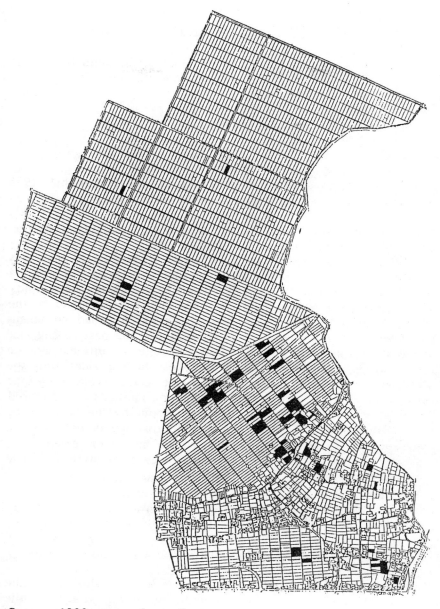

Source: 1983 survey by author.

Figure 5.4
Plot Locations of Rice Consumed by K Hamlet Farmers

Reorganization Project (K̄ochi Seiri Kumiai); (3) The Land Exchange of 1963 (Kōkan Bungō); and (4) The Land Improvement Project (Tochi Kairyō Jigyō) of 1980.

The 1933 Project to Change Vegetable Fields to Rice Paddies was a major shift towards monocropping rice. The project was directed at the type of crop rather than consolidation of land. The major goal of the 1951 Land Reorganization Project was reconstruction of paddies and field crop plots into ten are rectangular plots. Figure 5.5 illustrates paddy structure before and after the project. The project had minimal effect on merging of holdings but did change the actual size of the tilled paddy. Before the project, many subplots were required because of the natural slope of the land and the need to hold the paddy water level even. Due to the project less subplots were needed and tilled paddy size increased three-fold.

The Land Exchange of 1963 was relatively unsuccessful in its goal of land consolidation. The project was sponsored by the township government and was directed at holdings in the reclaimed marsh area. Only ten parcels belonging to residents of K Hamlet were exchanged.

The most recent project, described in Chapter 10, is the Land Improvement Project, which has been very successful in merging land in Asamizu and Takarae. As a result of the large amount of unregistered tenancy, however, there is some question whether the land is being consolidated for the tiller or the owner.

Uniformity of Land Quality

Improved soil conditions, irrigation and drainage due to the Land Improvement Project, along with the trend towards achieving suitability of the land for rice production, has tended to enhance the uniformity of land quality. Table 5.6 shows the land valuations used in the 1951 Land Reorganization Project.

A "1" was the highest rating possible based on a composite of such factors as water and drainage facilities, sunlight, soil conditions, and proximity to roads. After this table was constructed, the valuation system, while being based on the same general criteria, changed to a point system. The valuation is proportional to the number of points. Irrigation and drainage facilities seem to have been the key factor and carry more weight in the valuation because it was a limiting condition in production.

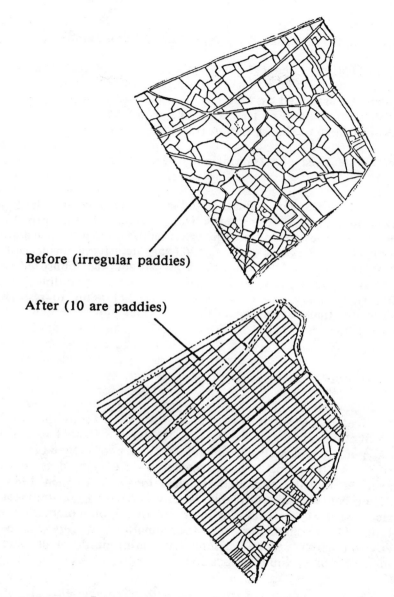

Before (irregular paddies)

After (10 are paddies)

Sources: Nakada-chō 1892, Nakada-chō Tochikairyōku 1983.

Figure 5.5
Paddy Configurations Before and After the 1951 Land Reorganization Project

Table 5.6
Land District Valuations in Nakada Township in 1951

Land District Category	Average Rating
Upstream irrigated district	2
District at the end of irrigation line	4
District beyond the irrigation line	6

Source: Nakada-chō Tochidaichō 1983.

By 1965 the valuation of the land in the three major environmental zones around K Hamlet ranged from a low of 1,547 points in the poor drainage areas of the reclaimed marsh to a high of 4,123 points in the area near the hamlet. Currently, the area near the hamlet also has high commercial value because its proximity to a main road makes it attractive for industry. The traditional paddy area was valued at approximately 2,800 points, showing that commercial value became more important than irrigation.

Uniformity of Land Use

Consolidation of land holdings has become much easier since a greater percentage of land has been converted to rice production. In 1955, farmers in K Hamlet planted 47.9 hectares (66 percent) of cultivable land (exclusive of vegetable plots) to rice. At the same time 6.5 hectares were planted to wheat, 6.4 hectares to oats, and 10.8 hectares to soybeans. In 1983 about 87 percent of all cultivable land was planted to single-cropped rice, with an occasional winter crop. Upon completion of the Land Improvement Project, the remaining differences between environmental zones should be minimized, and increased monocropping would be possible.

Government Policies Favoring Large-Scale Farming

The Land Commission of Nakada Township began implementing the Land Holdings Rationalization Project while I was engaged in fieldwork. Under this project, the Land

Commission (all farm land sales must be registered at the Land Commission) would give first buying option to those farmers who possessed above-average holdings for the township. By drawing concentric circles around each plot that was to be sold, the Land Commission contacted the nearest large farms to ask if the farmers wished to purchase the plot. It was, however, difficult for the Land Commission to force this method when the seller came to the office with a personally contacted buyer, which was often the case due to the minuscule amount of land that was available for sale. Land transferred under this program in Miyagi Prefecture is shown in Figure 5.6. According to Figure 5.6, large farms bought the land and very small ones sold it. Still, most of the buying and selling was between average-sized farms and so there has not been appreciable change of scale.

As mentioned earlier, the rate of land transfer is slow, and plans to merge plots through land transfer have failed to alter the fact that the average Nakada Township household owns land in seven or eight different paddy districts, with occasional non-contiguous paddies within them. The project with the most promise, the Land Improvement Project, has been rejected by Uwanuma. Plans to consolidate paddies without transferring the land (such as leasing and entrustment) are being adopted through the 1980 Act to Promote Agricultural Land Utilization (see Appendix A) and it is probable that for Nakada Township, leasing and entrustment may provide the key to the completion of the project without changing the ownership of the paddies themselves. Yet, only sixty-seven hectares of land were being leased through this program in 1985 and a fraction of that number was actually being used for consolidation of land holdings for the cultivator.

82

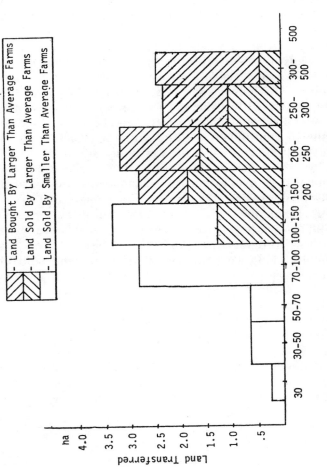

Source: Miyagi-ken Nakada-chō Nōgyō Iinkai 1983.

Figure 5.6
Land Sales for the Rationalization Project

NOTES

1. "Second wives" (<u>satogo</u>) and common-law wives (<u>naien</u>) have no inheritance rights (although their children do). Illegitimate children (<u>chakushi</u>) have one-half the rights of legitimate children. A fetus (<u>taiji</u>) has equal inheritance rights as a born child.

2. This law also provided an added incentive for landlords to lease their land unregistered. Besides being able to charge higher rents, they save on the inheritance tax because under the gift tax exclusion the heir is supposed to have had three years of farming experience. For these reasons some landlord farmers are leaving the legal records in their own name as if they were tilling the land by themselves. The idea of illegal leasing will be discussed in Chapters 7 and 10.

3. Legislation that has encouraged inheritance over time has lead to greater land fragmentation. Increased costs of production and desire to raise the standard of living pressure some farmers into selling their land before they die. Farmers now consider the possibility of mixing the inheritance (<u>sōzoku</u>) and gift (<u>zōyo</u>) methods to minimize inheritance taxation.

For example, part of the land might be given as a gift before death. In such a case a typical paddy actually worth two million yen might have an fixed property assessed valuation of four to six hundred thousand yen making it possible to give as a gift one paddy per year tax free. While the method of calculation is complex, there are a few simple rules. First, there was a gift tax deduction of six hundred thousand yen. Second, for taxable amounts up to two million yen the gift tax rate was thirty percent and the inheritance tax rate was ten percent. Third, for taxable amounts up to seventy million yen the gift tax rate was seventy percent while the inheritance tax rate was ten percent. Widows or widowers could deduct forty million yen or one-third of the inheritance, whichever is greater (Zenkoku Nōgyō Kaigisho 1982b).

In 1988 the exclusions for both the gift and inheritance taxes were increased. The gift tax exclusion was doubled from one hundred million yen to two hundred million yen. The following are the 1988 inheritance tax rates for <u>sōzoku</u> inheritance after the forty million yen exclusion (previously twenty million yen):

Tax Rate	Pre-1988 (in million yen)	1988 Tax Amendment (in million yen)
10%	20	40
15%	50	80
20%	90	140
25%	150	230
30%	230	350
35%	330	500
40%	480	700
45%	700	1,000
50%	1,000	1,500
55%	1,400	2,000
60%	1,800	2,500
65%	2,500	5,000
70%	5,000	over 5,000
75%	over 5,000	

Source: Nikei Daijiesuto September 20-21, 1988.

4. In 1982 seventy of the 131 cases of land sales were listed as due to types of "financial difficulties". "Labor shortage" accounted for thirteen cases and six cases were listed as due to the "cultivation inconvenience".

5. The difference would be even greater if the lowest end (Takarae Village) of the irrigation line were used, an area that also had the highest rate of absentee landlordism. The areas investigated for the districts were 14.6 hectares, 3.5 hectares, and 5.2 hectares, respectively, and all were used for growing rice. The Nakada land maps are exceptional since most land maps of the time do not list the land use or subplot number.

6

Land Reform

INTRODUCTION

This chapter describes the American land reform in Japan and its impact on Nakada Township. The low tenancy rate and increased economic power of Japan following the land reform has made it a model for other land reforms, and is an example of successful American foreign policy. The purpose of this chapter is not to debate whether or not the Japanese case should be a model, but rather to demonstrate some of the latent features of the land reform. The first part of the chapter makes clear the intent of the American land reform policy makers and the criteria they used in formulating the land reform, and distinguishes the Japanese case from the postwar land reforms in Korea and Taiwan. The second part of the chapter describes revisions in the land reform policy that make the Japanese case distinct from that of Korea and Taiwan, and the third part of the chapter deals with unforeseen local level problems caused by the reform.

THE AMERICAN LAND REFORM

The American land reform in Japan was based on American ideas on how to promote democracy. The occupation architects of the reform, Wolf Ladejinsky and Robert Fearey, held a common belief in Jeffersonian agrarianism, as did General Douglas MacArthur. Jeffersonian agrarianism emphasized the connection between ownership of private property and independence, liberty, and model citizenship. It also assumed the belief that private property was a natural right.[1]

The main goal of the reform, according to the December 9, 1945 SCAP Directive 411 on Rural Land Reform from General Douglas MacArthur to the Imperial Japanese Government, was "to insure that those who till the soil shall have a more equal opportunity to enjoy the fruits of their labor" (Walinsky 1977:579). This directive, based heavily on Ladejinsky's 1937 research on Japanese farm tenancy, was drafted to prevent agrarian unrest which might lead to fascism or communism. It listed five major "ills" which the land reform was to address :

1. Intense overcrowding of land.
2. Widespread tenancy under conditions highly unfavorable to tenants.
3. A heavy burden of farm indebtedness combined with high rates of interest on farm loans.
4. Government fiscal policies which discriminate against agriculture in favor of industry and trade.
5. Authoritative government control over farmers and farm organizations without regard for farmer interests.

The land reform came in two bills. The first, December 1945, was a watered-down version which was rejected by the Americans. The second bill affected more tenanted land and imposed more restrictions. The first bill abolished absentee landlordism, although it allowed non-absentee landlords a maximum tenanted land holding of five chō (about five hectares or 12.3 acres). The five chō limit would have had affected less than half of the tenant farmers and did not give tenant farmers and owner-cultivators control over the Land Commissions (Nōgyo Iinkai). This limit was lowered to one chō in the subsequent reform which took effect October 21, 1946.

In the final land reform bill (see Appendix A) all land owned by absentee landlords (fuzai jinushi) and all tenanted land owned by resident landlords in excess of four chō in Hokkaido and an average of one chō in the rest of Japan was available for purchase. Each prefecture had its own limits, which varied slightly according to local situations. Ownership of land leased out for cultivation could not exceed twelve hectares in Hokkaido and an average of three chō in the rest of Japan. Land Commissions were established to oversee land transfers and to control land rent. These commissions were comprised of five tenants (those owning less than one-third of the land they cultivated), three landlords (those cultivating less than one-third of the land they owned), and two owner farmers (who owned more than a third of the land they cultivated). All rents were to be paid in cash and were to be fixed at rates commensurate with the (low) price of land. The rent was not to exceed 25 percent of the total value of the crop for that year. Tenancy contracts were to be put in writing and their renewal could not be denied. Forest lands were not included in the reform (Hewes 1950). New buyers of agricultural land were to own a minimum of three tan (thirty ares). The reform was to be implemented within three years of the 1946 implementation and November 23, 1949 was established as the date for measuring the amount of tenancy (Chira 1982:96). Measuring the amount of tenancy was particularly necessary for determining the amount of land that tenants would receive from resident landlords.

There was considerable disagreement among the occupation planners regarding the upper limit on individual land ownership. On May 9th, 1946 Dr. W. MacMahon Ball, the British representative to the Allied Council, proposed setting an upper limit of three cho̱ on the total holdings an individual landowner could retain, and limiting to one cho̱ the amount of land a tenant could purchase (Chira 1982:95). Andrew Grad, of the Government Section of the occupation, was against the second part of MacMahon Ball's proposal which limited tenant purchases to one cho̱. This proposal, he reasoned, would have been difficult to administer and would have created many equal sized farms (Dore 1959:140). Grad's argument against MacMahon Ball's proposal was that farms differing in labor supply and extra-farm income opportunities would make farms of unequal size necessary. Farm size was not altered in the reform because MacMahon Ball's proposal to limit tenant purchases was not adopted.

Immediately after the land reform the Agricultural Land Act of 1952 restated the upper limit of land holdings at three hectares (see Appendix A). It also established a lower limit of three tan to qualify to buy land. More important, the Land Act prohibited agriculturally zoned land from being converted to non-agricultural use. Limits to tenanted land were set down where a ceiling of one hectare (four hectares in Hokkaido̱) was established for resident landlords and absentee landlordism was prohibited. Automatic legal renewal of leasing terms was guaranteed unless the tenant was notified at least six months to one year prior to the expiration of the contract. Tenant rights to cultivate land were to be inheritable. A ceiling for rents was also set up by the Land Commissions and rent was to be paid in kind.

Transferring to owner-cultivation most of the tenanted land in Japan, virtually abolishing absentee landlordism, and implementing its goals swiftly, thoroughly, and equally were achievements of the land reform. Achievement of the creation of small-scale Jeffersonian owner-cultivators who supported democratic principles resulted in strong postwar support of the farmers for the conservative party.

Tenanted land, reduced through the land reform, is shown by region in Table 6.1. Nationally, approximately 80 percent of tenanted land was affected by the reform. Thirty-six percent of the land purchased by the government through the land reform was owned by absentee landlords and 45 percent by resident landlords (Ogura 1982:746). Some areas such as Hokkaidō, Tōhoku, and Hokuriku were above the national average in tenanted land affected by the land reform. In 1950, after the reform, these three areas also had the lowest tenancy rates. Table 6.2 shows the prefectural limits established by the Land Reform. By 1979 the amount of land that was officially tenanted in Japan had decreased to less than 5 percent (Kondō 1981:367), and continued to be low even in the face of unregistered tenancy which gained popularity in the early 1980s.

Tenancy was dramatically reduced as shown in Figures 6.1, 6.2 and 6.3, which depict the change for one hamlet in Nakada Township. Figure 6.1 illustrates tenancy in the 1950s immediately after the Land Reform at which time there were no tenants in the former marsh area. Figure 6.2 shows the situation in 1961 when there were only a few tenants in the reclaimed marsh area and Figure 6.3, drawn from a fieldwork survey in 1983, depicts the rise of tenancy that remained unregistered with the Land Commission. The frequency of registered and unregistered tenancy was about equal and it was also clear that tenancy was occurring about equally in the three environmental zones.

In comparison with Korea and Taiwan, the Japanese land reform affected a much higher percentage of tenanted land because the political situation in Japan enabled a much more thorough redistribution. The landlord class was less affected in Korea and Taiwan, according to Olson (1974).

The Korean land reform began in 1950 and lasted until 1957, interrupted by the Korean War from 1950 through 1953. In Korea only 38.1 percent of tenanted land saw change through the land

Table 6.1
Regional Effect of the 1946 Land Reform in Japan

Region	Pre-Land Reform (On 11-23-45)		Land Reform Interim (11-23-45 to 8-1-50)		Post-Land Reform (On 8-1-50)
	Farm Land (in chō)	Tenanted Land (%)	Farm Land Affected (%)	Tenanted Land Affected (%)	Tenanted Farm Land Remaining (%)
Hokkaidō	725,887	48.7	47.5	93.1	6.7
Tōhoku	813,268	48.2	40.4	83.0	8.4
Kantō	873,961	50.6	39.5	77.5	12.5
Hokuriku	425,889	49.0	40.7	82.1	9.1
Tōsan	297,791	43.6	34.1	77.2	10.3
Tōkai	324,891	40.5	29.1	71.1	12.4
Kinki	352,315	44.9	33.5	73.2	13.6
Chūgoku	397,635	40.3	31.2	75.6	10.2
Shikoku	220,462	43.5	34.6	78.0	10.0
Kyūshū	705,597	41.0	30.0	73.2	11.0

Source: Nōchi Kaikaku Kiroku Iinkai 1951:614.

Table 6.2
Land Reform Upper Limits on Land Holdings by Prefecture

Prefecture	Maximum Acreage Leased out to Tenants by Resident Landlords	Maximum Acreage Leased and Owned
Hokkaidō	4.0	12.0
Aomori	1.5	4.5
Iwate	1.2	3.8
Miyagi	1.5	4.3
Akita	1.4	4.3
Yamagata	1.4	4.4
Fukushima	1.2	3.8
Ibaraki	1.2	3.7
Tochigi	1.3	3.9
Gumma	1.0	3.0
Saitama	1.0	3.0
Chiba	1.2	3.6
Tōkyō	0.7	2.2
Kanagawa	0.8	2.3
Niigata	1.1	3.6
Toyama	1.1	3.7
Ishikawa	0.9	2.7
Fukui	0.9	2.7
Yamanashi	0.7	2.1
Nagano	0.8	2.6
Gifu	0.7	2.2
Shizuoka	0.7	2.3
Aichi	0.8	2.4
Mie	0.8	2.4
Shiga	0.8	2.7
Kyōto	0.7	2.2
Ōsaka	0.6	1.9
Hyōgo	0.6	2.0
Nara	0.7	2.0
Wakayama	0.6	1.9
Tottori	0.9	2.6
Shimane	0.8	2.2
Okayama	0.7	2.3
Hiroshima	0.6	1.8
Yamaguchi	0.8	2.5

Table 6.2 (continued)

Tokushima	0.6	2.1
Kagawa	0.6	2.0
Ehime	0.7	2.2
Kōchi	0.8	2.1
Fukuoka	0.9	2.8
Saga	1.0	3.3
Nagasaki	0.8	2.3
Kumamoto	1.1	3.1
Ōita	0.7	2.3
Miyazaki	1.0	3.0
Kagoshima	0.8	2.2

Source: Nōchi Kaikaku Kiroku Iinkai 1951:230.

reform of 1950, although tenanted land comprised 60.4 percent of all cultivable land. As shown in Table 6.1, Japan had a lower percentage of tenanted land before the reform and a much higher rate of effectiveness as shown in the 1980 data in Table 6.3.

Before the land reform in Korea, the amount of tenanted land had decreased to 40.1 percent because of the combined effect of the threat from the north, the recent occurrence of the Japanese land reform, and landlord control of the agricultural policy. The Korean land reform was able to liberate land held by the Japanese. Between 1950 and 1957, 86.8 percent of the land held by the Tōyō Takushoku Company, a Japanese holding company, was redistributed by the land reform.

In Taiwan, the Land-to-the-Tiller Program took place in two stages and was designed primarily to help the nationalists, with whom the United States was aligned, achieve dominance over native landlords. The first stage involved selling public land, owned by Japanese settlers and corporations, to tenant farmers. The second stage redistributed land owned by (native) landlords in excess of three hectares. In other words, owner cultivation up to three hectares was permitted.

In 1945, 40.7 percent of Taiwanese farmland was tenanted. The land reform began in 1949 and reduced tenancy to 17 and 7.8 percent in 1955 and 1979 respectively (Mao 1982:735). In both Korea and Taiwan the landlords were also partially paid in industrial stocks (Olson 1974:60) as an inducement to participate in industrial development.

92

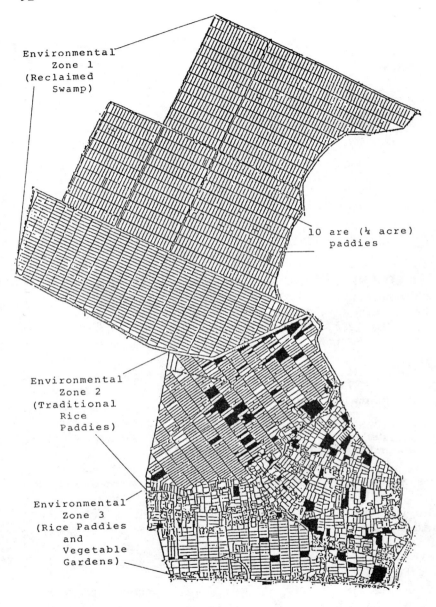

Environmental
Zone 1
(Reclaimed
Swamp)

10 are (¼ acre)
paddies

Environmental
Zone 2
(Traditional
Rice
Paddies)

Environmental
Zone 3
(Rice Paddies
and
Vegetable
Gardens)

Source: Nōrinsuisanshō Keizaikyoku Tōkei Hōkokubu 1955.

Figure 6.1
1955 Tenancy in K Hamlet

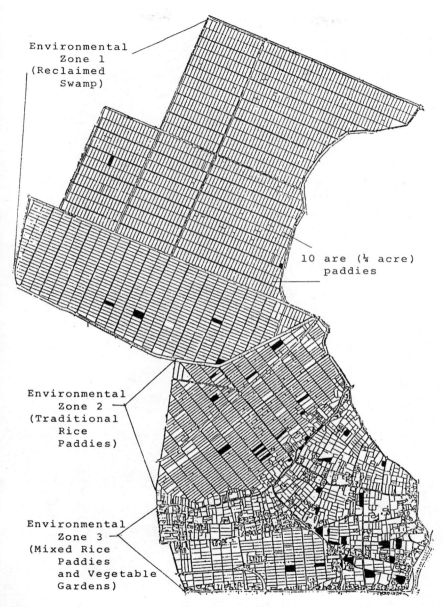

Environmental Zone 1 (Reclaimed Swamp)

10 are (¼ acre) paddies

Environmental Zone 2 (Traditional Rice Paddies)

Environmental Zone 3 (Mixed Rice Paddies and Vegetable Gardens)

Source: Data adapted from Nakadachō Nōgyō Iinkai 1961.

Figure 6.2
1961 Tenancy in K Hamlet

94

Environmental ——————
    Zone 1
(Reclaimed Swamp)

Environmental
    Zone 2
(Traditional
Rice Paddies)

Environmental
    Zone 3
(Rice Paddies
and Vegetable
Gardens) ————————

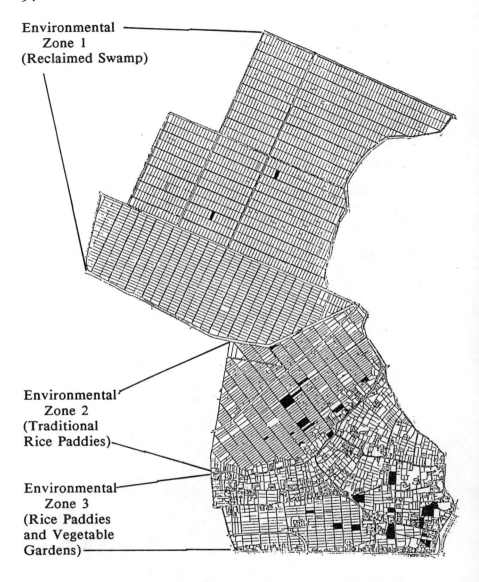

Note 1: Includes registered and unregistered tenancy.
Note 2: Actual plot locations have been changed within zones.
Source: Field interviews by author (1983).

Figure 6.3
1983 Tenancy in K Hamlet

Table 6.3
Tenancy Rates by Region in 1980

Region and Acreage	Rates of Tenancy on Cultivated Acreage			
	No Tenancy	Less than 30%	30 to 50%	over 50%
Japan				
2-2.5 ha.	74.2	19.0	4.4	2.4
2.5-3.0 ha.	72.6	19.4	5.2	2.8
3.0-5.0 ha.	69.2	19.9	6.5	4.3
over 5 ha.	61.5	18.8	9.5	10.1
Tōhoku				
3-5 ha.	80.4	14.8	3.1	1.7
over 5 ha.	70.7	17.3	7.0	4.9
Ninami Kyūshū				
3-5 ha.	50.9	25.2	13.2	10.6
over 5 ha.	44.3	22.5	14.7	18.5
Hokuriku				
3-5 ha.	46.4	34.3	13.3	6.0
over 5 ha.	29.2	26.3	21.7	22.8
Kinki				
3-5 ha.	47.5	18.6	12.8	21.0
over 5 ha.	34.0	16.3	12.1	37.5

Source: Isobe 1982:104.

The American occupation of Japan provided an unparalleled opportunity to use the land reform as an agent to promote rural conservatism in Japan. As noted in Chapter 8, the farmers of Japan exert pressure on the Liberal Democratic Party both through Nōkyō, the agricultural cooperative, and because they have greater voting power as a result of the rural district over-apportionment. From Table 6.4 we can see how the conservative vote in rural areas increased in the time frame immediately after the reform.

Table 6.4
Non-conservative Votes by Type of District 1947-1955

Type of District	Non-conservative Percentage of Total Votes			
	1947	1949	1952	1955
Villages (pop. less than 5,000)	47.5	34.6	28.6	30.4
Large villages and small towns (pop. 5,000-30,000)	47.8	36.5	29.4	31.4
Medium towns (pop. 30,000-150,000)	50.1	40.6	35.0	39.8
Large towns (pop. over 150,000)	53.6	46.1	44.1	44.9

Source: Cole, Totten, and Uehara 1966:410.

UNPLANNED EFFECTS OF THE JAPANESE LAND REFORM

As described thus far, the American land reform in Japan pursued the goals of promoting democracy through the creation of a large class of owner-farmers and decreasing absentee landowning[2]. To achieve this, land holdings were tallied and limits imposed on the amount of land which could be legally owned. This section of the chapter will describe several latent consequences of the land reform.

Farm Scale Unaffected

It is important to note, especially considering the 1987 United States' protest for free trade on agricultural products based on the cost inefficiency of Japanese farmers, that a small farm scale was not one of the "ills" at which the occupation land reform was directed. The land reform, by only focusing on quantity of land owned, did not noticeably change the average size of Japanese farms nor the percentage distribution of different farm sizes.

This is another key difference between the Japanese and Taiwanese land reforms. In Taiwan the number of farms under one hectare increased from 25 percent in 1952 to 35 percent in 1955 (Mao 1982:7356). As shown in Table 10.2 the biggest changes for Japan between 1908 and 1950 were only 3 to 4 percent in the three to five hectare category and the less than 0.2 hectare category of farm size. Furthermore, any subsequent change in farm scale has been minimal.

A legacy of the land reform's effectiveness at eliminating tenancy was the reluctance of farmers to create new leasing arrangements which might have enlarged the scale of farming. Even in 1986, farms over five hectares only averaged two hectares of leased land with the bulk of leasing being performed by farms less than two hectares as shown in Table 6.5. The average farm in Japan was 1.2 hectares (2.9 acres) of which 0.1 hectares were leased (Nihon Nōgyō Nenkan Kankōkai 1988:183). Ironically, the country which established the guidelines which sheltered small-scale farming in Japan was the same one to blame it for inefficiency forty-five years later.

Part-time Farming Remained the Most Viable Farming Strategy

Because the scale of farming remained small, the best way to increase farm household income was to take an off-farm job and continue farming part-time. It needs to be emphasized that part-time farming is not a post-land reform phenomenon. It has facilitated the rise of manufacturing from the early Meiji to the present. Osamu Aoki (1986:38) states that, according to data from twenty prefectures, part-time farm households constituted about 30 percent of the total farm households from 1884 to 1939. In 1939, 26.6 percent of farmers were tenants who did not own any land. These farmers were forced to seek jobs in industry to supplement their farm income.

Table 6.5
Distribution of Leasing by Size of Cultivated Land Area

Farm Size Managed (in hectares)	Percentage of Total Cultivable Farm Land		Percentage of Total Leased Farm Land	
	1960	1975	1960	1975
under 0.3	5.0	5.0	11.3	6.6
0.3 to 0.5	8.7	8.6	14.8	10.1
0.5 to 1.0	30.8	26.3	38.0	26.9
1.0 to 1.5	27.0	22.6	22.7	21.4
1.5 to 2.0	15.3	15.3	8.0	13.4
2.0 to 2.5	7.2	9.2	2.8	7.8
2.5 to 3.0	3.2	5.1	0.9	4.4
3.0 to 5.0	2.7	6.2	0.6	6.3
over 5.0	0.2	1.6	0.1	3.0
Total	100%	100%	100%	100%

Source: Isobe 1979:16.

The land reform temporarily reversed the trend and produced higher full-time farming rates until seasonal migratory job opportunities were created by the economic growth brought on by the business generated through supplying the United States military in the Korean War. The opportunity to own paddies coupled with the immediate rice shortage following World War II contributed to a resurgence of full-time farming. From 1942 to 1946 the rate of full-time farming rose throughout Japan from 41.9 percent to 55.4 percent. Accordingly, rates of part-time farming fell from 58.1 percent to 44.6 percent (Yoshida 1981:61) until they rebounded as shown in Table 1.5 in Chapter 1. The land reform planners did not envision the rapid postwar economic growth of Japan made possible by part-time farming. Prewar government fiscal policies discriminated against agriculture and instead favored industry and trade but nevertheless did so through achieving parity of agricultural and industrial incomes. This income parity and the small scale of production gave farmers the incentive to stay on the farm in order to combine on-farm and off-farm incomes in what would become a part-time farming phenomenon, the high degree of which distinguished Japan from its neighbors and the world.

Land Price Fluctuation

One major problem caused by the land reform was the low price of land. Viewed in retrospect the landlords were forced to sell their land at incredibly low prices. The purchase price of the land was calculated at 1945 prices (Dore 1959:139) and the average price for the typical rice paddy in 1945 was 8,970 yen. Although the price of a paddy was much higher than a decade before, the value of it was not. Rapid currency deflation meant a decrease in value so that by the time the landlords were paid, the amount they received was worth only one-tenth as much. The 1945 price was so low, in fact, that it was equal to less than half the value of the rice produced on the land. As Andrew Grad noted, the price of an average paddy in 1939 would have bought over three thousand cigarette packets or thirty-one tons of coal. In 1948, however, it only equalled thirteen cigarette packets or 0.24 tons of coal (Grad 1952:219).

In view of the rapid rise of land prices from the 1960s through the 1980s, the social reinterpretation of the land reform by both landlords and former tenants has taken on new meaning on what they gained or lost in terms of the land's present-ommercial value. In most cases, the land reform Land Commissions, who were required to base the allocation of land on the best interest of the self-cultivating farmers, gave former tenants the best agricultural land of resident landlords whose holdings exceeded the prefectural limits on land leasing. This left resident landlords with land which had the least agricultural productive value. In some cases this land later turned out to have the most commercial potential.

Increased Fragmentation

Increased fragmentation as a result of the land reform was due to several factors. These factors concerned the redistribution of land held by resident landlords, who were allowed to keep part of their land as self-cultivators, but who owned land in excess of the prefectural limit. In the First Land Reform of 1945 it was the right of the landlord to choose which paddies he would keep. However, in the more severe Second Land Reform of 1946, landlords lost this right. The responsiblity for choosing which paddies would remain those of the landlord was delegated to the Land Commissions. Because most landlords had more than one tenant, this became a complex situation. Local Land Commissions also felt morally responsible to leave the landlord with some vegetable and paddy land rather than leave him with one or the other because both were necessary for subsistence.

Moreover, the problem was how to redistribute the land of a resident landlord who leased non-contiguous paddies to two or more different tenants. For example, if landlord X owned one hectare in excess of the prefectural limit but had leased one hectare each to tenants A and B, problems arose over how to divide the one hectare equally to A and B since neither could receive their full one hectare (because resident landlords were allowed to keep up to the prefectural limit). Figure 6.4 illustrates a common method of division used in such cases. Following these guidelines, the landlord household retained two separate paddies while his former tenants A and B each ended up with fifty are parcels. In the process, two new paddies were created.

Pre-reform Resident Landlord-Tenant Plot Distribution

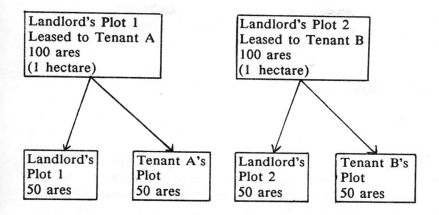

Post-reform Self-cultivator Plot Distribution

Figure 6.4
Land Fragmentation Resulting from the Division of Resident Landlord Holdings to Multiple Tenants

Quality and Quantity

As noted above, the land reform put upper limits on the quantity of land cultivated and leased out. Because quality of land was incorporated into the price, there was an incentive for resident landlords to keep the most valuable land. In Cherry Tree Village, landlords abandoned reclaimed swamp paddies and retained the more highly valued land closer to the hamlet.

Areas with higher rates of absentee landlordism typically experienced more land fragmentation after the reform. Poor farmers, treasuring each parcel of land, sold off paddies one by one when they became indebted. For example, the comparison of Uwanuma District (which had a high degree of prewar resident landlord control) and Takarae District (which had a high degree of prewar absentee landlordism) in Chapter 5 reveals the fragmenting effect of prewar poverty.

Under conditions of equal slope, smaller paddies provided more versatility in land use. Small paddy sizes also allowed the farmer more control during water shortages so that some of the paddy could be diverted to non-paddy use. Water levels in the paddy could be held more constant and farmers could use ecological variations in soil, wind, and sun conditions to plan effective use of various seed varieties adapted to specific situations.

Repatriates from the War

During the war many Japanese families left their farms in the care of tenants or relatives while they migrated to Japanese territory in Korea or China to farm or serve in the Japanese Occupation. In fact, at the time of the first Korean land reform over 11 percent of the land in the southern part and 4 percent of the land in the northern part of Korea were owned by Japanese. The return of Japanese from Korea and Manchuria presented a problem for the reform because in most cases their move to the colonies was recorded in the household registers. Thus, it was possible for Land Commissions to regard them as absentee landlords if friends or relatives claimed tenancy. When the land reform was implemented, there were some cases where friends or relatives caring for the land were classified as tenants and thereby able to purchase the land. Therefore, it was common for some rural area returnees who lost their land in this way to migrate elsewhere upon their return.

Kin Alliances and Tenancy Claims

In the case above, some close kin relatives made tenancy claims against repatriate relatives who had moved to the colonies during the war. In other cases, close kin relatives avoided tenancy claims in order to preserve kin alliances. The last was occasionally true in the Tōhoku area where hierarchical arrangements of main and branch households often incorporated informal tenancy relations. These informal tenancy relations were recognized because in the process of creating branch households, main households usually had the branch household work on the land for a stipulated period before allocating land to them. The main households were obligated to assist the branch in the process of heir and spouse recruitment which were necessary for succession. The mutual bond based on both material and ancestral relations sometimes took priority over tenancy claims.

REVISIONS IN THE LAND REFORM

The land reform in Japan differs from the land reforms in Korea and Taiwan in that most of the original provisions in the Japanese case have been overturned. Major revisions as documented in Nōgyō Roppō (Nōrinsuisanshō 1983) are the 1961 Agricultural Basic Law, the 1962, 1970, and 1980 Amendments to the Agricultural Land Act of 1952, and the 1965 Reward for Cooperation in the Land Reform Act (see Appendix A).

The 1961 Agricultural Basic Law was the first comprehensive policy for agriculture in postwar Japan. The law attempted to enhance farming efficiency by increasing scale while shifting the surplus rural population from farming to industry. By encouraging some rural people to give up farming, it was hoped that plots would be consolidated and farms made larger. The rural population did move into secondary and tertiary sectors but, as shown in Table 10.2, did not appreciably change the scale or land holding pattern. Farmers did not sell their holdings because the law promised income parity between the agricultural and industrial sectors. Instead of a mass permanent migration of land owners to the city in search of better opportunities, farmers journeyed to the city only during the off-season to supplement their farm income. During the 1970s, as rural industrialization occurred, farmers slowed their seasonal migrations because job opportunities were available in their local communities.

The 1962, 1970, and 1980 amendments to the Agricultural Land Act of 1952 were perhaps the most significant changes in the land reform. The Agricultural Land Act of 1952 established a "land to the tillers" policy, with the Land Commissions acting as regulators of leasing agreements, rent rates, land transfers, and upper land holding limits on leasing and cultivating. The 1962 amendment gave some private corporations the right to exceed the upper limits on agricultural land ownership. The 1970 amendment was the most important alteration because the "land to the tillers" policy was modified to "land for the purpose of efficient use in agriculture," the upper limits on ownership were abolished, and some cases of absentee landownership were permitted. Perhaps the most significant part of this amendment was that it denied tenants the right to compensation if their landlords decided to sell the land they worked. Land tenanted before this date carried the right of the tenant, which in much of the postwar period was equal to approximately one-half the value of the land. Therefore, this amendment, along with substantial compensation given to prewar landlords in 1965 (see Appendix A), can be seen as a victory for the prewar landlords.

CONCLUSION

While the land reform was effective in terms of the goals it set out to accomplish, the upper limits left Japan with an agricultural sector that had such a small-scale that international competition in many products was nearly impossible. After 1961 agricultural policy in Japan came to favor a larger, more "efficient" scale of production with no special interest in tenancy or ownership issues, which contradicted the land reform's priority for small-scale owner-cultivators. After the 1970 amendment to the Agricultural Land Act, farmers who sensed the government's change in policy direction often neglected to register their tenants with the Land Commissions in order to avoid legal documentation and regulation of tenancy. This became a "gray area" because in 1980 a third amendment to the 1952 Agricultural Land Law exempted participants in the 1980 Act to Promote Agricultural Land Utilization (see Appendix A) from registering their land with the Land Commissions. One of the main goals of the program was the leasing of agricultural land.

Caught in the wave of changing agricultural policy, the Land Commissions themselves took on a new role of facilitating land transfer to larger farms whenever a parcel of land became available for sale. They also continued to play a critical role in promoting farm succession through the 1970 Farmers Pension Act (see Appendix A), which supplements the national retirement social security system. In order to receive the farmers pension, a farmer was required to transfer the land to an heir before retiring at the age of sixty.

The land reform did accomplish most of its stated goals. Unfortunately, the planners failed to foresee the needs of the 1980s and 1990s when foreign demands of free trade necessitated changes in the basic rationality of production. How the land reform will be evaluated in the future will depend on the on-going social relations governing land ownership and land use as well as whether or not agricultural land is valued for its agricultural productive potential or commercial real estate value.

NOTES

1. See Susan Chira's (1982) outstanding account of the philosophy held by the occupation planners.

2. Interested readers should consult <u>Land Reform in Japan</u> (1959) by Ronald Dore and <u>Nōchi Kaikaku Tenmatsu Gaiyō</u> (1951), published by the Ministry of Agriculture, for a fuller description of the background and implementation of the reform.

7

Technical Change and the Social Effects of Mechanization

The Rice Ballad

Taking apart the character for rice (米),
You will find it will take eighty-eight
(八 丶 十 丿 丶) hands every time.
Offering one kernel of rice as a meal,
Is nothing to be ashamed of.
We are descended from rice.
Our parents are rice.

(Tōhoku folksong)[1]

THE CAUSES OF MECHANIZATION

Japanese farmers are sometimes referred to as "rentai peasants". "Rentai" literally means "connected holidays". It refers to the two vacation periods, one in the spring and one in the fall, when the national holidays line up close enough together that an industrial worker can, depending on which date the weekend falls, have four or five days off in a row without using up any vacation time. The spring official holidays consist of the Showa Emperor's birthday called "Green Day" (April 29), Constitution Day (May 3), and Children's Day (May 5). By taking as few as one or two days off, it is possible to gain up to five days off and when the weekends are properly aligned with the holidays, it is possible to have as many as ten vacation days in a row by only taking off three work days. If part-time farmers effectively use the "rentai" vacation for the labor intensive spring planting period they can produce an entire rice crop by doing the rest of the necessary work on the regular weekends.

Rural industrial development, while providing the jobs required for farmers to maintain their households at an income parity with the urban areas, has nevertheless subordinated agriculture to industry. In their efforts to mechanize, farmers have focused on how to shorten the labor time necessary for growing rice so that the time can be better invested in earning higher wages in industry. Both mechanization and the technological development of which it is part are inseparable from the social processes which create them. Similarly, their impact is social in nature.

Agricultural mechanization, as a type of technology, has reduced the amount of human labor necessary in agriculture. Comparing labor reduction and yield increase, the former can be singled out as the main goal in devising new technology in the postwar era. In the late 1950s rototillers were introduced while rice transplanters, reaper-binders, and harvesters became popular in the late 1960s through the mid-1970s. Taken together these machines are responsible for the remarkable reduction of the amount of labor necessary to grow ten ares of rice. This number has fallen from 155 hours in 1955 to a low of 56.5 hours in 1983 (Nihon Nōgyō Nenkan Kankōkai 1955-1986). Yield increases during the same period were only up 30 percent and remain below the five hundred kilogram per ten are ceiling.

It is, however, essential to keep the idea of mechanization separate from other types of technology. In the Tōhoku Region the greatest gains in crop yield have resulted from improved crop varieties, better irrigation and drainage facilities, chemical fertilizers, insecticides, herbicides, and fungicides. In particular, improved crop varieties that respond well to fertilizer have increased yields and have shortened the growing season to reduce the frost risk. Table 7.1 shows the crop varieties developed during this century in Aomori Prefecture, one of the six Tōhoku Region prefectures. The relation between nitrogen utilization and yields is shown.

Nitrogen utilization to raise crop yield was a major factor in the 1960s advance of the Sasanishiki variety of rice in Miyagi and Yamagata Prefectures, the principal rice-growing belt in the Tōhoku Region. In K Hamlet there has been a rapid increase in rice yields in the postwar period. According to informants, in the early postwar period most farmers raised the Fukubozu variety which yielded five to six bags per ten ares. The next development was the Sasashigure variety which yielded seven to eight bags. In 1983 over 95 percent of farmers raised the Sasanishiki variety, which has an average yield of eight to nine bags.

Table 7.2 demonstrates how the practice of applying nitrogen side-dressing affects the relative yield of the Sasanishiki variety of rice. In Miyagi Prefecture the average yield is sixty kilograms lower than in Yamagata where a second top-dressing is usually applied. One of the reasons why Miyagi prefecture farmers do not use as much top dressing as those in Yamagata Prefecture is the extra time required.

Table 7.1
The Development of Rice Varieties, Nitrogen Utilization, and Crop Yields in Tōhoku

Rice Crop Variety	Period	Nitrogen Utilization (kg./10 ares)	Crop Yield (per 10 ares)
Kaminoo	1915-1935	6.388	420
Rikuha 132	1935-1955	6.75-7.5	480-495
Norin 1	1935-1955	9.08	480-540
Norin 17	1950-1955	9.08	550
Fujisaka 5	1950-1960	10.58	600
Towada	1955-1965	10.58	620
Fujiminori	1955-1970	11.58	630
Reimei	1965-1975	12.08	640
Akihikari	1975-1985	n.d.	n.d.

Sources: Usami:1981:19; Miyagi-ken Nōgyō Kaigi 1982:175.

Table 7.2
Fertilizer Top Dressing in Miyagi and Yamagata Prefectures

Prefecture	No Top Dressing	One Top Dressing	Two Top Dressings
Miyagi	45%	47%	8%
Yamagata	2%	36%	62%

Source: Usami 1981:19.

Having completed fieldwork in the rice belts of both areas, my general impression is that the Shōnai Plains farmers (Yamagata Prefecture rice belt) and their Nōkyō organization are modern farming oriented. This may be due to a one hectare larger scale of production in the Shōnai Plains and because part-time farming opportunities are greater in Miyagi Prefecture than in the Shōnai Plains area.

This study agrees with the observations of Ishino and Donoghue (1964) who comment on the rapidity of technological diffusion in rural Japan. There seems to be a "reticular formation" in which there is a close network between farmers, township agricultural section workers, agricultural extension workers[2], land commissions, agricultural experimental station technicians, agricultural cooperative guidance workers, agricultural high schools, university professors, agricultural equipment sellers who sponsor rural seminars, and the land improvement districts which regulate water and administer the Land Improvement Projects to enlarge the paddies and improve the irrigation system. It is also true as Penelope Francks (1984) has documented in the Saga Plain case that Japanese industrial development policy has had a profound effect on the type of agricultural innovations.

Unlike other forms of technology, which have mainly focused on yield increases, mechanization has served the purpose of enabling households to allocate more of its members' time to wage labor instead of agricultural labor and also obviated the need for hiring outside help. The postwar emphasis on mechanization to save labor costs is made clear in Table 7.3. The average yield and amount of time necessary to till ten ares demonstrates that while yield remained relatively constant in the postwar period, labor was saved by the introduction of machinery. Table 7.3 also reveals the rise in the cost of labor and machinery. In an effort to keep labor costs down, farmers reduced their labor time by more than half during the twenty-five year period investigated.

In order to clearly demonstrate the process of mechanization to save labor through the aggregation of labor tasks, I have included in Table 7.4 a 1937 distribution of labor tasks by month. At that time rice cultivation was much more labor intensive and depended entirely on draft animal plowing and human labor.

It should be noted that pre-mechanized Japanese agriculture was more labor utilizing because the farm household's labor was spread over the entire year (with the exception of January and

Table 7.3
Labor and Machinery Costs of Rice Production, Time and Yield/
10 Ares

Year	Machinery Costs (yen)	Labor Costs (yen)	Labor Cost/ Machinery & Draft Animal Cost (yen)	L a b o r Time/ 10 ares (hours)	Yield per 10 ares (kg.)
1955	716	5,648	73.8	154.8	435
1956	749	5,707	72.6	153.4	472
1957	835	5,871	72.6	150.8	423
1958	950	6,754	73.6	156.7	401
1959	885	5,734	70.5	140.2	440
1960	1,190	6,123	69.7	143.9	486
1961	1,654	7,572	70.6	147.5	476
1962	1,944	8,195	71.7	132.4	469
1963	2,006	9,971	78.4	128.9	463
1964	2,486	11,767	80.1	125.3	425
1965	2,825	12,903	80.7	123.4	506
1966	2,869	14,681	82.4	122.7	461
1967	3,794	17,399	81.5	127.6	518
1968	4,323	18,226	80.6	119.4	518
1969	5,379	19,679	78.4	113.7	505
1970	6,343	19,808	75.7	104.6	534
1971	7,118	20,846	74.5	101.0	486
1972	8,135	18,804	69.8	83.3	525
1973	10,560	20,922	66.4	77.4	511
1974	12,055	26,571	69.4	74.2	469
1975	14,619	31,022	67.9	71.3	558
1976	18,043	36,476	66.9	70.6	480
1977	20,313	36,570	64.3	65.6	500
1978	22,648	40,189	63.9	66.9	557
1979	24,421	40,173	62.1	63.6	526
1980	25,885	39,785	60.5	59.6	407

Source: MAFF Statistics quoted in Miyagi-ken Nōgyō Kaigi 1982:9.

Table 7.4
Monthly Distribution of Labor Used in Rice Agriculture in 1937 (in persons per month)

Operation	1	2	3	4	5	6	7	8	9	10	11	12	Total People	% of Total
Preparing Seedbed			0.6	0.4	0.1								1.1	4.2%
				1.0									1.0	3.9
First Plowing					3.0								3.0	11.7
Second Plowing						2.0							2.0	7.8
Leveling						2.3							2.3	8.9
Transplanting						2.0	2.5						4.5	17.5
Weeding								0.3	0.3				0.6	2.4
Water Maintenance							0.3						0.3	1.2
Disease and Pests					0.6	0.5	0.2						1.3	5.1
Fertilizer										3.6			3.6	14.0
Cutting											2.0		2.0	7.8
Binding											1.0		1.0	3.9
Drying											1.0	1.0	2.0	7.8
Preparation Bagging												1.0	1.0	3.9
Total Persons			0.6	1.4	3.7	6.8	3.0	0.3	0.3	3.6	4.0	2.0	25.7	100.0%
Total Percent			2.3	5.4	14.4	26.4	11.7	1.2	1.2	14.0	15.6	7.8	100%	

Source: Yoshida 1981:102.

February). This strategy was also used by the Russian peasants studied by Chayanov (1966). Postwar mechanization has been labor displacing. In Table 7.5 we see the 1980 situation with respect to production operations.

Tables 7.4 and 7.5 give evidence that steps which formerly took several months to complete in 1937 were not only reduced in the actual labor hours to accomplish them but also were aggregated into a shorter period of the calendar year. In 1937 the beginning stages of rice culture (exclusive of raising seedlings) comprised 32.3 percent of total labor time, and being in separate stages farmers were able to spread the tasks over a three month period. In 1983, however, plowing was usually done only once and most often occurred within a month of the actual transplanting. Figure 7.1 demonstrates that only a relatively small number of paddies were plowed in the winter or early spring well in advance of spring transplanting. In 1983 spring plowing was usually included as part of "transplanting" with the autumn or winter plowing stage omitted. Thus, the transplanting stage now combines the plowing, preparation, fertilizing, paddy preparation, and transplanting stages, which taken together represent 27.6 percent of the total labor time.

Table 7.5
1980 Human Labor in Tōhoku Rice Agriculture

Operation	Labor Time (hours per 10 ares)	Percent of Total
Seedling Boxes	0.8	1.2%
Raising Seedlings	8.2	12.7
Plowing and Preparation	6.2	9.6
Fertilizing and Paddy Preparation	3.2	4.9
Transplanting	8.5	13.1
Post-planting Fertilizing	1.1	1.7
Weeding	5.8	8.0
Water Maintenance	11.0	17.0
Disease/Pest Management	1.4	2.2
Harvesting	15.5	23.5
Threshing	3.3	5.1
Total Time	64.7	100.0

Source: Tōhoku Nōseikyoku: 1980:148.

114

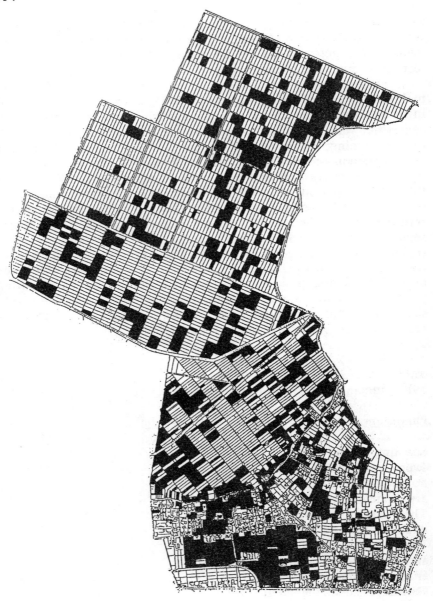

Source: Moore 1985:199.

Figure 7.1
Distribution of Plots Plowed Before Spring Transplanting

The autumn harvest is another peak labor period and represented 28.6 percent of total labor time in 1980. As the scale of Japanese agriculture is increased, there will be pressure to reduce this peak labor period through new technology.

Nevertheless, many farmers still take advantage of their paddies being in different environmental zones. If a farmer has little equipment and scattered paddies in zones that vary with respect to water availability, human labor for one operation such as transplanting can be spread out over a longer period of time.

As seen in Table 7.6, there appear to be three distinct stages of postwar mechanization in Nakada Township. The rototiller stage, which came first, began in 1960 and rapidly spread until tractors with over twenty horsepower began replacing rototillers in 1973. The tractor stage has never really replaced the rototillers both in Nakada Township and the country. In 1987 in Japan there were 2,682,210 rototillers and 1,904,070 tractors, with over half of the tractors being less than twenty horsepower (Nihon Nōgyō Nenkan Kankokai 1988:477).

The second stage of mechanization, characterized by small-scale motorized equipment, began in the late 1960s when the reaper-binder started to gain acceptance. These binders were popular because in addition to cutting the rice stalks, they tied small bundles which could be easily stacked to dry. Between 1973 and 1975 the number of transplanters increased three and a half times. By 1977 rototillers, transplanters, and reaper-binders were diffused to almost one-half of the farming households. The harvester was used by over one-third of the households.

The third stage is characterized by larger machinery both in size and horsepower. In the early 1980s the number of tractors over twenty horsepower rose to about one-quarter of the farming households. At the same time the number of combine and power grain driers rose to 6 percent and 14 percent, respectively. These machines still have not gained widespread acceptance due to their cost and impracticality on small plots.

Large-scale farmers who have been able to consolidate plots, and farmers in areas where the Land Improvement Project has enlarged paddies to thirty ares, have been the first to purchase combines and dryers. Drawbacks to the new dryers are insufficient storage capacity and therefore lower drying efficiency. Even larger dryers have only a storage capacity equivalent to the yield of four paddies (ten are paddies), while most farmers work three times this acreage. This necessitates quick drying of rice to accommodate a new batch, and according

Table 7.6
Nakada Township Agricultural Mechanization

Year	Tractors over 20 h.p.	Roto-tillers	Trans-planters	Reaper-Binder	Harvesters	Combine	Dryers
1960		185					
1961		388					
1962		556					
1963		771					
1964							
1965		706					
1966		1,212					
1967		1,388		12			
1968		1,566		89			
1969			5				
1970		1,516	10	232		5	
1971		1,473	33	447	55	6	
1972		1,398	115	623	97	12	
1973	54	1,530	285	851	181	17	
1974							
1975	152	1,552	999	1,194	468	40	163
1976	211	1,665	1,259	1,315	656	67	
1977	269	1,692	1,371	1,394	891	76	166
1978						96	
1979	383	1,688	1,436	1,444	1,065		224
1980			1,522				
1981	515	1,733	1,552	1,528	1,255	135	
1982	718	2,098	1,720	1,599	1,365	171	345
1983						180	390
1984							
1985	1,156	2,107	1,590	1,590		266	465

Source: Nōrinsuisanshō Keizaikyoku Tōkei Hōkokubu 1960-1986.
Note: In some cases the numbers of tractors and roto-tillers have been added.

to many, the quick drying method causes uneven drying of kernels, producing poor tasting rice.

Non-mechanical forms of new technology have also been aimed primarily at saving labor time. One example is the milky herbicide used to kill weeds in the paddy. As noted before, weeding represented 17.5 percent of total labor in 1937 and in 1983 it represented 9 percent of a much lower total labor amount. A second example is the amount of time necessary to manage the water flow in the paddy. In 1937 this was 2.3 percent of the total labor. In the 1950s and 1960s the time necessary to control and maintain water increased from eighteen to twenty-two hours per ten are paddy due to the labor needed to maintain the mid-season drying out period (nakaboshi) which is from June 27 to July 1. The current Land Improvement Project will dramatically reduce the amount of time necessary for water maintenance by sustaining improved water lines and drainage to and from each paddy.

THREE LARGE-SCALE FARMING ATTEMPTS

The Tadano Farm

Founded in 1957 by Naosuke Tadano, Tadano Farms was the largest privately owned family farm in Japan. It was made possible because Naosuke Tadano purchased for reclamation a large tract of flood zone river bottom land along the Hasama and Furukawa Rivers in Nakada Township. The eleven branches of the Tadano family farmed and managed the land owned by Naosuke Tadano. This amounted to 25.5, 7.5, and 4.5 hectares of land in Nakada, Tajiri and Nango Townships respectively.

Tadano Naosuke was head of a branch household that in 1920 was established with only 0.4 hectares of land. Through buying and selling vegetable fields and flood lands next to rivers, Tadano became one of the largest self-made landlords by amassing over seventy hectares of reclaimed rice field, fifty hectares of unreclaimed land, and had 170 tenants on thirty-six hectares (Kawai 1971:61) by the end of the war. The post-war land reform aimed at abolishing absentee landlordism took away most of Tadano's land even though he had fallowed much of the land at the time of the reform and was not leasing it out. The land reform set an upper limit of 4.5 hectares for Miyagi

Prefecture but allowed Tadano to keep nine hectares in Tajiri Township because there was a potential for family farming.

In 1957 Naosuke Tadano bought large tracts of land in Nakada and Nangō Townships with the goal of introducing large-scale machinery to be able to produce rice with a low cost production. The paddy sizes varied between one and three hectares in size (Kawai 1971:57). In 1970 the Tadano average labor time per ten ares gained national attention because it was only 30.7 hours compared to the prefectural average of over 100 hours (Tadano 1970).

Three major factors, however, lead to the dissolution of the farm. The first problem was the tremendous cost of reclamation. The reclamation project took five years, from 1957 to 1963, and cost 36 million yen. Naosuke Tadano borrowed from Nōkyo, Shinren (Federated Trust Bank), Nōrinchūkin, and private banks at annual interest rates as high as 14 to 24 percent. The second problem was that the newly reclaimed land had low yields and suffered from frequent flooding. Being in a flood zone, Tadano had difficulty getting crop flood damage compensation and government cooperation in constructing flood control measures. As Kawai (1971:62) noted, these conditions produced insufficient income to repay the loans and was the main cause of dissolution. The situation was somewhat alleviated by the rice curtailment policy of 1970 and 1971. During this time Tadano was able to take advantage of the subsidies for fallowing his land and therefore did not increase his debt.

The third problem for the Tadano Farms was inheritance and management of the farms after Naosuke Tadano's death. In 1975 the land was inherited as described in Table 7.7.

Not only was ownership fragmented among surviving relatives who lived in different locations, but also the land of many of these relatives was split between the three locations of the Tadano Farms.

A public corporation had to be set up in order to repay the large debt that the farm owed, further complicating the matter. The nine hectares in Tajiri Township were left intact because they were fully owned, but the Nakada and Nango Township farms were organized under a public corporation so that the land could be leased out. Because the land often flooded and because the people, including some of the relatives listed below, who leased it invested money in improvement, a second lien was put on the property by those leasing it (Ōizumi 1981:116). Therefore, some of the relatives have a second lien while others do not.

Table 7.7
Tadano Farm Inheritance

	Land Owned (in hectares)
Main Household	
Second Eldest Son	11.0
Branch Households	
Third Eldest Son	6.5
Fourth Eldest Son	6.5
Fifth Eldest Son's	
Eldest Son	0.5
Second Eldest Son	0.5
Eldest Daughter	2.5
Second Eldest Daughter	0
Other Relatives	
Sister	4.5
Niece	2.5
Nephew	2.5

In 1987 under government pressure to promote large-scale farming, the Land Commission was trying to figure out a way to salvage the farm by consolidating the fragmented holdings.

Nakada Township Large-Scale Farming Experiment

No doubt influenced by the early successes and attention drawn to the Tadano Farm, Nakada Township has always been interested in large-scale farming. In fact, during this time of excitement with large-scale farming, three of the farmers sold their farms and migrated to Hachirogata Land Reclamation Project to be pioneers in the largest scale public sponsored project in Japan. In 1967 at a township sponsored construction deliberation meeting in which participants discussed the future of Nakada farming, Nakada Township decided to experiment with large-scale farming under the guidance of and with aid from the Agricultural Policy Research Station.

Fourteen adjacent landowners, led by a prewar owner-cultivator who had 2.9 hectares, were helped in forming the "Large-scale Mechanization Experimental Rice Field Cooperative." This cooperative enabled the township to collectively utilize 3.4 hectares of paddy land so that it could be farmed using large-scale equipment from 1970 to 1974. Two 1.5 hectare blocks were formed, each containing eight paddies restructured out of the thirty-eight paddies owned by the fourteen landowners.

The experiment, which used large-scale tractors and combines, resulted in a significant reduction of labor with only 6.6 hours per ten ares needed compared with the prefectural average of 104 hours. However, most people were discouraged by the inconvenience, low quality of rice, and high costs associated with the large equipment. The machines continually broke down, got stuck in the mud, and produced only third and fourth grade rice because of the chipped and unevenly dried kernels. The failure of the project and particularly poor performance of the combines is no doubt one of the reasons why the township is presently well behind the prefectural average for the use of mechanical combines. However, more important, the project was doomed by the dramatic advent of the gentan (see Chapter 10) government policy restricting rice production.

The Township-Cooperative Joint Helicopter Spraying Project

In 1970 the township and two of the four agricultural cooperatives started contracting three helicopters to spray 496 hectares of paddies for rice blast disease (imochibyo). The cost was very minimal and a representative of each of the local Nōkyō was responsible for checking the accuracy of the spraying in his hamlet based on the observation of selected hidden pieces of black plastic upon which the white speckles of the milky spray would land. If a piece of black plastic had not been sprayed, then the area would be re-sprayed.

This method of spraying was very effective for the control of rice blast disease and soon spread to the control of other diseases. The method was both cost effective and labor reducing. Prior to this a farmer would have to team up with other farmers to spray an area. If the owner of the adjacent paddy did not control the disease, then the disease could spread.

THE SOCIAL EFFECTS OF MECHANIZATION

There are eight principal social effects of mechanization. These include: (1) an increase in part-time farming and decrease in hired farm labor; (2) a decrease in traditional cooperative labor groups; (3) new types of cooperation to save labor; (4) a change in the temporal division of labor; (5) a change in the division of labor; (6) a change in production goals and use of environmental zones; (7) a change in the sexual and age division of labor; (8) and a change in rice symbolism.

An Increase in Part-time Farming and Decrease in Hired Farm Labor

Part-time farming increased in Nakada Township between 1955 and 1980 while the number of full-time farmers dropped from 55.6 percent to 10.0 percent over the same time period. Part-time farmers category II (more than 50 percent of farm household income derived from off-farm sources) have grown from 15.9 percent in 1955 to 58.4 percent in 1980. Remarkably, the number of farming households has remained at twenty-nine hundred.

Because rural industrial development and subcontracting increased during the late 1960s, local employment was provided and the continuance of the household structure insured. Nakada Township closely approximates the national average in its part-time farming rates and demonstrates the effectiveness of Japan's co-development policy of agricultural mechanization with rural industrial development. If rural industrial opportunities had not been present, it is very likely that either the rate of tiller-transplanter adoption would have been slower or that people would have left their rural communities because of an inability to secure heirs due to younger persons moving to the city.

The amount of hired labor on the farm has also decreased and the high cost of this labor has no doubt been a driving force behind rapid mechanization. As a result of mechanization, the total amount of agricultural day labor dropped from 124,313 days in 1960 to 89,401 days in 1971, although the number of farm households employing that labor grew from 1,775 to 1,899. Part of this change was also due to the fact that industrial wages were higher than agricultural wages during the expansion of the Japanese economy.

Decrease in Cooperative Labor

Cooperative agricultural labor in Nakada Township includes two types of *yui* labor between individual households and one pan-hamlet type involving all landowning households. The first type of yui was cooperative labor between a main household and its branch. In Nakada Township most main households were also prewar landlords, and branch households, sometimes lacking sufficient land, would work for the main household which had land but not enough people to till it. Because the land reform limited the amount of land a landlord could rent out, many of the branch households received land from their main household at the time of the reform. The labor obligations resulting from using main household land or being indebted to main households rapidly decreased with the land reform.

By 1950 most of the first type of yui cooperative labor had ended, with the number of households doing tetsudai labor testifying to this. Tetsudai labor is one household "helping" another household and differs from roryoku kokan which literally means "labor exchange". Tetsudai and roryoku kokan labor relations are more informal and less permanent than were the yui relations of the past. In 1960 the number of households doing tetsudai was 836 (total of 7,520 labor days) and by 1971 had declined to 319 households (total of only 2,220 labor days).

The second type of cooperative yui labor was the labor exchange between households as equals. Sometimes several households would work together to perform a task that could be more efficiently completed by a group than an individual. This was particularly true for peak labor periods in rice agriculture. However, unlike the first type of yui which rapidly declined, the second type became more common. At the height of this type of cooperative labor exchange in 1970, 1,151 (39 percent) of the households exchanged a total of 25,224 days of work. This represented a 191 percent increase over a ten year period during which the rototiller and transplanter were becoming popular (Nakada-chō 1955-1983) and characterized a trend towards a less hierarchical hamlet social structure.

The third type of cooperative labor is on the pan-hamlet level. An example of this type of labor is the semi-annual irrigation drainage ditch cleaning. This is called eharai in local dialect and is shown in Photo 7.1. The ditches are located in the open paddy area surrounding several districts (aza). The open area is named after the hamlet and the hamlet members have a form of usufruct right to till the land. Other persons can buy land in these districts, but it is acknowledged that it is K Hamlet's domain since it is immediately adjacent to the hamlet, and member households have traditionally tilled these plots. Each household that owns land is required to send one representative to assist in the effort.

Recently, non-farm job obligations or personal commitments have occasionally interfered with a household being able to send their representative to the communal work session on the designated day (usually a Sunday). Thus, a fee system has been instituted for those who do not participate, the rationale being that the hamlet would have to pay the township to do the work if it was not done by the households.

Some of the hamlet members would prefer a system based on a per paddy fee rather than on equal household participation and are opposed to the present system which is based on usufruct rights rather than ownership rights. The Township Agricultural Land Commission favors a policy which would enable the ditch cleaning to be a service for which the fee would be based on land ownership and carried out by the township rather than the hamlet.

New Types of Cooperation Developed from the Need to Save Labor and Cut Costs

New ways to cooperate arose out of the farmer's shared need to reduce agricultural labor time in favor of time spent in non-farm work. For example, helicopter spraying for rice blast disease was coordinated by the Uwanuma Nokyo and the township office. During the summer the fields were sprayed seven times, and as described earlier, each hamlet sent a representative to verify that all paddies were accurately sprayed. New types of cooperation, however, are usually compensated rather than strictly voluntary on a non-paid basis and there seems to be a tendency for farmers to pay according to acreage.

Also, large machinery is more often shared than is small-scale machinery. For example, in the over-thirty horsepower tractor category in 1982, forty tractors (55 percent) were shared compared to thirty-three which were privately owned. This compares to 147 (23 percent) shared and 498 privately owned in the twenty to thirty horsepower range. Seventeen percent of combines, another large piece of costly equipment, are shared. Small machinery, such as rice transplanters, were shared only 12 percent of the time in 1982.

A Temporal Change in the Division of Labor

Prior to the war, each spring at the beginning of rice transplanting season, the farmers of K Hamlet would look towards Mount Kurikoma to see if the melting pattern of the snow made the shape of a school of fish. When the pattern was just right, it was time to plant. Planting time came earlier than it does now and the autumn harvest continued longer than today. In fact, because the harvest work lasted into January it was customary to celebrate New Years, Shōgatsu, in February.

In 1962 rototiller mechanization made it possible to complete work earlier and celebrate New Years on January 1 as did city-dwellers. At the same time, All Soul's Day, Obon, was switched from July 13-15 to August 13-15 because farmers were busy in mid-July but had a lull in mid-to-late August shortly before the September harvest. As shown previously, agricultural work in 1937 was spread out over most of the year and this utilized household labor more effectively. Growing winter crops also served to utilize household labor during the off-season for rice cultivations.

In the postwar period, rapid mechanization occurred in areas where labor was traditionally most intensive such as in the transplanting and harvesting seasons. Polyculture was transformed into monoculture and micro-environmental zones that differed in soil types, shade, form, and irrigation and drainage were converted into paddies with identical growing conditions. Many vegetable crops as well as winter crops were virtually discontinued during this period so that the fields could be changed into new paddies (kaiden). Because more rice paddies of similar growing conditions were formed, periods of peak labor demand were created, providing the stimulus for technological innovations such as the rice transplanter. Gradually, as new machines freed up more labor time, diverse tasks could be accomplished within a shorter period of time. With transplanting, for example, the four stages shown in Figure 7.2 would have taken several months to perform in 1937. The data from Figure 7.2 were collected over a two day period (May 8th and 9th, 1983) and most farmers had completed transplanting within a two week period.

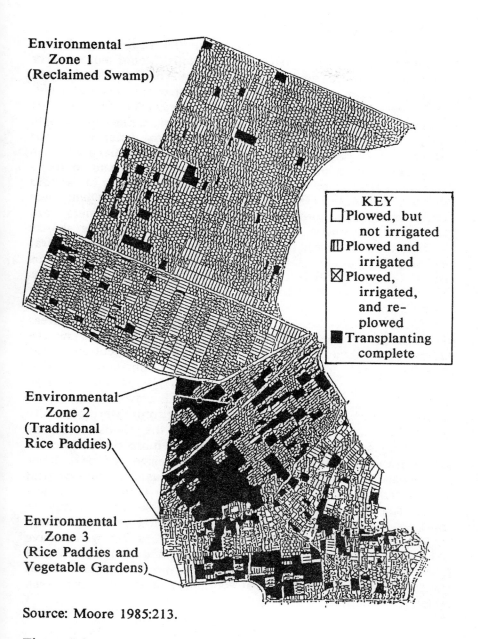

Environmental
    Zone 1
(Reclaimed Swamp)

KEY
☐ Plowed, but
    not irrigated
⊞ Plowed and
    irrigated
⊠ Plowed,
    irrigated,
    and re-
    plowed
■ Transplanting
    complete

Environmental
    Zone 2
(Traditional
Rice Paddies)

Environmental
    Zone 3
(Rice Paddies and
Vegetable Gardens)

Source: Moore 1985:213.

Figure 7.2
Rice Transplanting Stages and Environmental Zones

The Division of Labor

The four distinct environmental zones described in Chapter 2 provided an opportunity to spread household labor more evenly over a longer period of time. The diverse environmental zones provided varying access to the irrigation water necessary for transplanting, and these zones played an important role in the era of pre-mechanization. Each zone also had different drainage, with the former marsh area continuing to be the last area to be planted due to poor drainage[3]. Traditionally each work group could complete the transplanting process in one zone and move to the next when water or soil conditions permitted. In Nakada Township, scheduled spring plowing was sometimes delayed in the former marsh area because it takes several weeks to dry out after a rain. Water availability is also more reliable in the traditional paddy area (honden). Although the environmental dissimilarities between the zones are disappearing, they are still visible due to the differences in water drainage and irrigation. The darker areas of Figure 7.2 delineate the final stages of transplanting and the lighter areas show the beginning stages of transplanting. Completion of the Land Improvement Project will likely improve the possibilities for introducing large-scale equipment that would enable farmers to further save time in the planting process.

Change in Production Goals and Use of Environmental Zones

How and where rice is produced is affected by mechanization. In 1987 farmers ate rice primarily from the traditional rice paddy area because it was better tasting and they avoided eating rice from the reclaimed marsh because it was poorer tasting. Rice production methods differed in these two zones. For instance, rice produced for the farmers own consumption (as opposed to rice they sell to the government through Nōkyō) was sun-dried in bundles attached to cross-supports tied onto poles that were driven perpendicular into the paddies. Since the bundles on the top of the pole dried faster than those on the bottom, the poles had to be reversed to dry the rice evenly, which was accomplished by taking the top bundles from one pole and placing them on the bottom of an empty pole. As each pole was cleared, it received the top bundles from the next pole. Stacking rice to dry is shown in Photos 7.2 and 7.3.

Sun-dried rice or machine-dried rice generally met the government's definition of first grade rice. Therefore, rice grown in the reclaimed marsh area and sold to the government was usually machine dried, allowing valuable human labor to be allocated elsewhere. Thus, rice drying techniques in the reclaimed marsh area differed from those of other areas. This distinction is important because the average farm household distributed approximately 120-180 kilograms of rice to relatives each year in addition to that eaten by the farm household. As mechanized production methods become more widespread, especially with the increased use of the combine and power dryer, sun dried rice will become even rarer.

Many attempts have been made to produce homogeneous environmental zones in order to produce higher rice yields. Traditionally the farmers of Nakada Township have had a close relationship with their land, and its "living" quality. Soil can be classified as "skinny" or "fat" (tochi ga yaseru, tochi ga futoru) and surrounding mountains such as Mount Kurikoma were watched to detect the fish pattern of melted snow which signaled the beginning of spring planting.

Historically the lives of the peasants in this area were also intricately related to the water level in the Kitakami River, which forms one of the borders of the township. In some years the melting snow and spring or autumn rains would bring floods. In other years when there was a shortage of rainfall, the level of the river dropped and there were water shortages that resulted in great famines. In 1947 and 1955 Nakada Township endured great floods which resulted in the river overflowing onto the rice fields causing crop damage and destruction. However, along with the floods came several inches of river bottom silt that caused a noticeable increase in the next year's rice productivity. As a result, farmers began supplementing their land with "guest soil" (kyakudo) from the surrounding hills, mountain tops, and river bottoms, realizing that it could be beneficial.

In 1964, using interest free financing from the Agricultural Improvement Fund and some of their own financial resources, the Cherry Tree Village Soil Tillage and Cultivation Cooperative (Sakuraba Kōdō Baiyō Kumiai) was formed with 712 members (Kōho Nakada 1966). The cooperative's main purpose was to upgrade the soil in the traditional rice growing zone adjacent to the former marsh area and in the reclaimed marsh area itself.

The members applied fifteen tons of mountain soil per ten are paddy which resulted in a 10 percent increase in yield. Using carts and horses, each year the cooperative members applied kyakudo to between four and five hectares. After 1970 the cooperative was discontinued because the reclaimed marsh began to produce top grades of rice. Nevertheless, many farmers still apply guest soil once every three to four years.

Change in the Sex and Age Division of Labor

New technology has affected both the sex and age distribution of labor, illustrated by Tables 7.8 and 7.9. Women traditionally transplanted and weeded the rice and in a sense these same women have themselves become transplanted into subcontracting factory jobs as described in Chapter 9. There is a tendency for men to drive the new machines such as rototillers, tractors, and combines. Mechanization has affected the age distribution of labor by allowing older people to continue working in the fields. Older people, however, preferred to use more labor intensive techniques rather than the newer machinery favored by their heirs. For example, the older generation ordinarily used straw rope to tie rice bundles onto the pole while the younger generation, many of whom cannot make straw rope both because the rice straw has become shorter and because they do not know the technique, has adopted nylon rope. The young heirs typically choose the nylon rope because it will save time.

More than one prospective heir has been lured into staying in or marrying into the community by the promise of new agricultural equipment. Table 7.9 shows that only the young are learning to use the new technologies. The young age of the

Table 7.8
1955 Division of Labor by Sex and Age in K Hamlet

Average Male Age	34.7 years
Average Female Age	28.31 years
Male Percentage of Total Labor	46.1 percent
Female Percentage of Total Labor	53.9 percent

Source: Nōrinsuisanshō Keizaikyoku Tōkei Hōkokubu 1955.

Table 7.9
Distribution of Farm Labor by Age and Sex in 1983
(for rice only)

Farm Task	Sex	Number	% of Total	Average Age
First Fertilizing	M	23	74.2	58.6
	F	3	26.8	55.9
Growing Seedlings	M	51	54.8	48.3
	F	42	45.2	46.3
Plowing	M	38	95.0	47.2
	F	2	5.0	38.0
Transplanting Machine	M	41	89.1	47.9
	F	5	10.9	55.0
Transplanting Helper	M	33	47.8	45.8
	F	36	53.2	45.8
Disease and Insect	M	36	61.0	50.9
Spraying	F	23	39.0	47.1
Water Management	M	55	70.9	55.1
	F	16	29.1	55.1
Second Fertilizing	M	36	85.7	54.7
	F	6	14.3	50.3
Putting Up Posts	M	45	84.9	49.1
	F	8	15.1	46.0
Harvester Binder	M	40	87.0	48.9
	F	6	13.0	44.2
Hand Harvesting	M	21	42.9	50.9
(used with combine)	F	28	57.1	49.7
Combine	M	6	100.0	41.5
	F	0	0.0	----
Mechanical Drier	M	6	85.7	46.5
	F	1	14.3	52.0
Drying on Posts	M	61	51.7	46.9
	F	57	48.3	43.0
Field Transport	M	43	76.8	46.2
	F	13	23.2	48.2
Threshing	M	56	60.9	45.3
	F	36	39.1	41.8
Marketing	M	53	72.7	49.0
	F	19	26.4	47.6

Source: Personal field interviews.

combine drivers contrasts with the age of laborers in other categories. However, the older generation claim to have a "feel" for the fine-tuned operations such as putting on the top dressing fertilizers, tsuihi, and regulating the water. Tsuihi is applied by hand, and the plants irrigated, when color and growth is appropriate. Some members of the older generation wonder how the younger generation will apply the fertilizers at the correct time since they lack knowledge about the proper timing and "feel" for the life cycle of rice plants. Perhaps the younger generation will be spared from water regulation by the automated system planned for the new Land Improvement Project.

Another age related change in production is that young children are used less in harvesting. They are too young to drive the machinery and such tasks as weeding that were required prior to the introduction of the machines and chemicals have been eliminated. The elementary school children do, as shown in Photo 7.4, still participate in the rice harvest by collecting locusts (inago) in the paddies in school drives. These locusts are sold to a company which makes cooked sugary delicacies out of them (but most people save a few to cook for their own consumption).

Changes in Rice Symbolism

Rice is the most dominant symbol of Japan. Even the eighth century Kojiki and Nihonshoki mention that Japan is the "Land of Abundant Rice". In Nakada Township rice is still offered to the gods on New Years morning. Likewise, many of the customs in which rice is used are representative of cooperative social values of the people that produced it. Ueage is a rice transplanting rite in which the farmers pray to Ebisu, god of good fortune. Ueage consists of up-rooting a few stalks of recently transplanted rice and offering them to Ebisu by placing them on the Shinto altar (kamidana). Ueage also used to be associated with the work group taking a vacation after the transplanting was complete. In the Tōhoku Region this vacation is called sanaburi. Currently, only a few families go together on an outing for sanaburi. The same situation prevails at the harvest celebration, niwabarai, when the farmers thank Ebisu. Formerly it was an occasion for great celebration but no longer is important. The decreased observance of traditional customary rites associated with agriculture seems to be associated with

Photo 7.1
Cooperative Pan-hamlet Ditch Cleaning

Photo 7.2
Support Posts for Sun-drying Rice

Photo 7.3
Stacking Rice for Sun-drying

Photo 7.4
Collecting Locusts for the School Drive

large-scale agriculture and the introduction of machinery which replaces human labor.

LARGER SCALE FARMS BUY THE NEW MACHINERY AND SUPPORT LARGE-SCALE FARMING PROJECTS

K Hamlet farms which were above-average in size bought new combines, large tractors, transplanters, and driers more often than smaller farms. Once a farmer had acquired larger machines and the latest equipment, he was more likely to support ventures such as the Land Improvement Project, which will provide the infrastructure leading to more efficient use of large-scale equipment.

Larger equipment can exploit new time-efficient advancements, such as the current trend towards smaller transplants. The older method (chūbyō) used sprouts which were about sixteen centimeters long while the newer method (chibyō) uses sprouts which are about ten to eleven centimeters long (Miyagi-ken Nōgyō Kaigi 1981:28). By 1981, 32.9 percent of Nakada Township farmers were employing the newer method (Miyagi-ken Kishō Kasai Hasama Chihō Taisaku Kaigi 1982:33), one of the advantages being that transplants can be raised in greenhouses under controlled conditions. Capitalizing on this, the agricultural cooperative was developing new centers to relieve the farmers of the burden of raising their own transplants. By 1986 the percentage of chibyō had risen to 47 percent (Miyagi-ken Hasama Nōgyō Kairyō Fukyūsho 1987:19). Smaller transplants also have the advantage that more can be carried by the machine, and many new machines can only utilize small transplants. One drawback of this new technology is the short length of the seedlings, which can be easily submerged in water if the paddy is uneven or even if a strong wind shifts the water from one end of the paddy to the other.

CONCLUSION

As governmental policies favoring large-scale agriculture make the new technology seem "more rational" because it saves labor, the labor utilizing advantages of small-scale agriculture are forgotten. This situation has a direct effect on the cohesiveness of hamlet life in rural Tōhoku. Small-scale agriculture was adaptive because the household and hamlet could regulate production. The household could, for instance, modify the number of persons in its labor force either through reproduction or by recruitment of new members. In addition, small-scale agriculture favored effective utilization of different environmental zones.

The future of large-scale agriculture is in the hands of government planners and Nōkyō bankers. If agriculture becomes more capital-intensive, costs of production and loan interest rates will become key factors. Even the increased use of space-saving biotechnology (which employs genetically improved hybrids grown in electronically controlled greenhouses) is based on large capital investments. Both government policy favoring large-scale farming and the considerable amount of money necessary to farm, cause it to be increasingly important to lower the costs of production.

If these policies are successful, there can be no doubt that the future of hamlet level agricultural cooperation is doomed and there will be no need for the cooperative paddy labor that is already rapidly disappearing. With hamlet cohesiveness reduced, it is probable that the interests of the one or two full-time farmers in the hamlet would be different than that of the majority of the households. The latter would become full-time workers forced from their land by the new policies. If the land were entrusted, however, rather than sold to full-time farmers, it might be possible to retain an agricultural group identity and even experience increased cooperation based on shared ideas concerning preservation and maintenance of hamlet land.

NOTES

1.

Komebushi

Kome to iu ji o bunseki surebayo
Hachi ju hachi tabi no te ga kakaru.
Okome hitotsu mo somatsu ni naranu
Kome wa warera no oya ja mono.

Kome no naru ki de tsukuri shi waraji yo.
Fumeba, koban no ato ga tsuku.
Kane no naru ki ga nai to wa uso yo.
Shinbō suru ki ni kane ga naru.

The first verse emphasizes the labor intensive aspect of rice production. (Eighty-eight hands refers to the forty-four persons necessary to do the work required in just one paddy). The second verse describes how rice is a "tree of money."

From the tree that bears rice also comes straw sandals,
The footprints of which resemble the old gold coins.
It's a lie that there is no tree which bears money,
For perseverance is a tree that will bear money.

2. The fundamental role of the agricultural extension service cannot be minimized. In 1982 the list of activities it promoted in Nakada Township was substantial. It included thirty-eight separately funded projects ranging from paddy reorganization to programs aimed at specific techniques and diseases. Special projects aimed at certain age groups (such as the Senior Citizens Beef Feeding Project and Project to Foster Farm Successors) were also included (Miyagi-ken Hasama Nōgyō Kairyō Fukyūsho 1982). The extension service also keeps very close tabs on the township rates of diffusion for new machinery. The rates are published in the regional extension service publications (Miyagi-ken Kishō Saigai Hasama Chihō Taisaku Kaigi 1981). The Land Commissions, whose primary concern is supervising agricultural land tenure, also annually publish literature concerning the diffusion of specific rice growing techniques. Typically these publications focus on seed varieties, disease prevention, planting techniques, and yield. Last, the role of the agricultural cooperatives is important in technological diffusion. For instance, comparing the four agricultural cooperatives in Nakada Township the diffusion rate for mechanical dryers varied from 73 percent to 17 percent.

3. When I completed this survey, two large tractors were stuck in the mud in the old marsh area.

8

Nōkyō:
The Agricultural Cooperative

INTRODUCTION

Nōkyō is the private cooperative institution through which farmers market their rice and other agricultural products and which serves as the major financial institution in rural Japan. Agricultural cooperatives in Japan were legislated from above shortly before the turn of the century but have a grass-roots appeal that is reflected in their continued popularity among farmers. Nōkyō serves as a rallying point for farmers to work out technical agricultural problems as well as solve social problems such as the bride shortage described in Chapter 9.

The political significance of Nōkyō in rural Japan rivals that of the township office. When major political figures arrive in town, the leader of Nōkyō usually follows the mayor in the perfunctory welcoming speeches. The reason for the political clout of Nōkyō is its control over the flow of rice and money. Rural signs of the 1960s encouraging farmers to "put their savings in Nōkyō" ("Chokin wa Nōkyō e") has lead Nōrinchūkin (The Agricultural Central Bank of Japan), the quasi-governmental financial arm of Nōkyō, to be ranked sixth in deposits among the world's largest banks (Nash 1988) twenty-five years later. This tremendous growth is based on the financial powers legislated to Nōkyō through the American Occupation. At the same time protecting its financial control over rice and the flow of government rice subsidies, Nōkyō has stood firm representing the farmers against free trade on rice demands of the United States Rice Millers Association.

Nevertheless, both farmers and pro-free trade industrialists have recently questioned the degree to which Nōkyō represents the farmers. Critics also question the degree of local association autonomy and assert that decisions are made more for the benefit of the institution than its constituent members. These critics point to increased farm land foreclosures on debts farmers have made through Nōkyō when buying modern farm machinery. This chapter describes the development of agricultural cooperatives in Japan, Nōkyō's political power base of rice and money, its role vis-a-vis the township government and the land improvement district, and its role in promoting collective social activities.

THE DEVELOPMENT OF AGRICULTURAL COOPERATIVES IN JAPAN

The development of agricultural cooperatives in Japan has followed five distinctive stages as follows[1]:

Phase One (1900-1911):   Formation of Industrial Cooperatives

Phase Two (1911-1925):   Expansion of the Cooperative System

Phase Three (1925-1943):   Completion of the National Cooperative and Wartime Control

Phase Four (1947-1965):   Growth through Financial Control over the Rice Marketing Network

Phase Five (1965-present):   Amalgamation and Financial Expansion and Diversification

Phase One: In the 1890s farm income comprised nearly one-third the national income. The formation of industrial cooperatives in Japan in the late nineteenth century was a result of government efforts to curb the declining economic position of farmers in the growing capitalist economy. Influenced by Prussian agricultural cooperatives while studying abroad, Yajiro Shinagawa and Tōsuke Hirata pushed for legislation to promote agricultural cooperatives in Japan.

In 1899 the Agricultural Association Law was passed establishing cooperatives for farmers who had paid at least two yen of land tax or landlords possessing over forty ares (0.98 acres) of land. Although the cooperatives were prohibited from buying and selling, they did contribute to the improvement of agricultural techniques.

In 1900 the Industrial Cooperative Association Law was passed creating credit, purchasing, production, utilization, and marketing agricultural cooperatives. Marketing cooperatives, the dominant type of cooperative, were formed mainly for selling rice, wheat, barley, and cocoons. It was common for landlords to dominate the formation of cooperatives (Midoro 1982:203, Ogura 1982:308).

Phase Two: Following World War I, the number of cooperatives in Japan rapidly doubled following the formation of a nationwide marketing system. By 1925 there were 12,880 credit cooperatives, 10,924 purchasing cooperatives, and 8,226 marketing cooperatives. This increase in cooperatives is shown in Table 8.1. As a result of the spread of these cooperatives, the number of cooperative federations linking the individual cooperatives also increased. In 1910 there were thirteen federations, and by 1925 there were two hundred.

The 1913 Rice Tariff and 1921 Rice Law were attempts by the Japanese government to protect the domestic rice market from the rising colonial imports of Taiwan and Korea. The Rice Tariff imposed a duty on the colonial rice while the Rice Law initiated public management of the rice inspection system as well as government control of rice pricing. Subsidies to control imports were granted through the Agricultural Warehouse Law.

Before 1917 the credit cooperatives were cooperatives geared toward obtaining loans. After 1917, however, they were allowed to also receive savings from members. In 1923 the Agricultural Central Bank was formed as the central credit organ for agricultural cooperatives and their federations.

Phase Three: The period 1912 through 1926 is characterized by a large number of tenant and labor disputes which contributed to the establishment of the tenants rights described in Chapter 4. The government, sensitive to the growing class antagonism and social upheaval in rural communities, exacerbated by the world depression and uneven regional development resulting from the growth of the munitions industry, initiated loan programs to save farmers from going bankrupt. These included the 1932 Farm Rescue Diet, the 1932 Rural Rehabilitation Program, 1933 Debt Clearance Unions Act, and the 1938 Farmland Adjustment Act (Appendix A). The 1932 Rural Rehabilitation Program, in

Table 8.1
Development of Industrial Cooperatives in Japan from 1900 to 1946

Year	Number of Coops	Credit Coops	Marketing Coops	Purchasing Coops	Utilization Coops	Warehousing Coops	Urban Credit Coops
1900	21	13	5	7	4		
1905	1,671	986	344	492	178		
1910	7,308	5,331	2,904	4,242	908		
1915	11,509	9,738	5,110	7,257	1,674		
1920	13,422	11,901	7,032	9,821	2,448	687	65
1925	14,517	12,880	8,226	10,924	4,358	1,740	224
1930	14,082	12,104	8,336	10,292	5,576	2,568	259
1932	14,352	12,211	8,360	11,042	6,184	2,986	267
1934	14,815	12,678	11,120	12,108	8,792	3,686	271
1936	15,457	13,433	12,846	13,249	11,299	4,665	269
1938	15,328	13,538	13,642	13,784	12,794	5,135	280
1940	15,229	13,430	13,563	13,742	13,126	5,550	282
1942	14,235	12,971	12,541	13,187	12,777		

Source: Midoro 1978:205

particular, had great impact on the numerical growth and widening of the structure of the cooperatives through increasing the membership of tenant farmers.[2] This growth was also accomplished through promoting multipurpose cooperatives and more extension work. In fact, the number of technical staff rose from 6,708 technicians in 1920 to 14,624 in 1939 (Ogura 1982:736). New programs also initiated rural hamlet level (burakukai) and sub-hamlet level cooperative organizations (jikko kumiai). In urban areas people were organized into neighborhood associations called chōnaikai.

After the China Incident of 1937, many of the individual farm associations (nōkai), their activities, and their federations were amalgamated into agricultural associations (nōgyōkai) and brought under the control of the government. This took its final form in 1943 when the Agricultural Organization Law was passed joining the industrial (sangyōkai) and farm associations into one system. In 1942 the Food Control Act was passed regulating the storage and distribution of rice, which was in short supply. The domestic self-sufficiency of rice in 1942 had fallen to an all time low of 78.6 percent (Ogura 1982:702).

Phase Four: On December 9, 1945 the occupation forces abolished the old cooperative system and directed the government to establish voluntary Rochdale-type democratic cooperatives which would be free from landlord domination. In 1947 the Agricultural Cooperatives Act was passed granting Nōkyō the right to perform banking functions. This in itself might have seemed innocuous, but combined with Nōkyō's domination over the flow of rice and government subsidies for rice, this act presented an opportunity unequaled since the Edo Period. In the Edo Period, the Osaka rice dealers, later to become large trading companies, were the exclusive dealers in trading rice for money for the feudal lords. After the Agricultural Cooperatives Act, Nōkyō was able to turn its financial influence into political influence particularly in regard to policies favoring increases in the rice price. Increases in the rice price not only meant that farmers earned a higher income, it also meant that more money was flowing through the Nōkyō financial pipeline.

Phase Five: In 1961 the Law for the Promotion of Agricultural Cooperative Amalgamation was passed. Thereafter the number of Nōkyō associations steadily declined as is shown in Table 8.2. While the increased size of the cooperative has aided its response

Table 8.2
The Decline in the Number of General Purpose Cooperatives

Year	Number of Coops	Year	Number of Coops
1949	13,308	1969	6,470
1950	13,314	1970	6,185
1951	13,300	1971	6,049
1952	13,341	1972	5,688
1953	13,311	1973	5,488
1954	13,191	1974	5,198
1955	13,154	1975	4,942
1956	12,335	1976	4,803
1957	12,704	1977	4,763
1958	12,572	1978	4,657
1959	12,406	1979	4,583
1960	12,221	1980	4,546
1961	12,050	1981	4,528
1962	11,586	1982	4,473
1963	10,813	1983	4,373
1964	10,083	1984	4,317
1965	9,135	1985	4,303
1966	7,320	1986	4,267
1967	7,209		
1968	7,074		

Source: Midoro 1982:211; Zenkoku Nōgyō Kyōdō Kumiai Chūōkai
1987a:433.

to member demands, some complain that the organization is too big.

Despite the numerical decline in cooperatives, membership has grown. In 1960, cooperatives counted 5,780,000 full individual members and 756,000 associate members (Ogura 1982:732). Despite the fact that the number of farms in Japan has rapidly declined, in 1984 the number of full members was nearly the same at 5,548,075 and the number of associate members increased to 2,392,214 persons (Zenkoku Nōgyō Kyōdō Kumiai Chūōkai 1987a:435). Although each cooperative association has some latitude in defining "full" and "associate" membership, usually a full member tills over ten ares and works more than ninety days per year at farming. Associate members must live within the territory of the cooperative and can utilize the cooperative services but cannot take part in decision-making at the annual meeting of members.

While membership in Nōkyō has remained constant, the number of cooperative staff members has rapidly increased. In 1960 the number of staff members stood at 180,000, compared to 295,000 in 1984. If all employees of the prefectural and national economic federations and coordinating groups of Nōkyō are counted, they number over 400,000. Remarkably, only 18,983 are specifically designated as farm management agents (einō shidōin) (Zenkoku Nōgyō Kyōdō Kumiai Chūōkai 1987a:437). In the general purpose cooperatives, 28 percent of the employees work in the financial banking area (Nihon Nōgyō Nenkan Kankōkai 1988:334).

The structure of Nōkyō is shown in Figure 8.1. Particular attention should be drawn to the three levels of Nōkyō: the local level, the prefectural level, and the national level. The Nōkyō Chūōkai "Central Committee" was established in the 1954 revision of the Nōkyō Law and is funded 94 percent by member dues (Nihon Nōgyō Nenkan Kankōkai 1983:304). There is one national central committee and forty-seven prefectural central committees under it. The main roles of the Chūōkai are leading the organization, supervising projects, educating the members, and dealing with the media. The Chūōkai also sets goals and participates in long term planning. The chairman of the Chūōkai was a prominent member in the drafting of the major agricultural planning directive of Japan called "The Basic Direction of Agricultural Policy for the 80s" (see Appendix A).

Each hamlet or sub-hamlet has its own jikkō kumiai whose function is mainly decided by the local group. In K Hamlet there are two such jikkō kumiai, roughly constituting two neighborhood areas of the hamlet. Each of these jikkō kumiai elects its own officers, calculates its production yields, figures the percentage of each grade of rice produced, and discusses production problems. Likewise, each sub-hamlet cooperative takes a trip to a hot spa in January. The K Hamlet jikkō kumiai typically visit Narugo Spa, which is a resort about one hour away. The leadership of the sub-hamlet level group seems to be in the hands of the younger farmers with the highest level of agricultural expertise. The former leader worked for the land improvement district and the new leader was an agricultural extension agent. This new leader owns a personal computer and provides printouts with production statistics for the members.

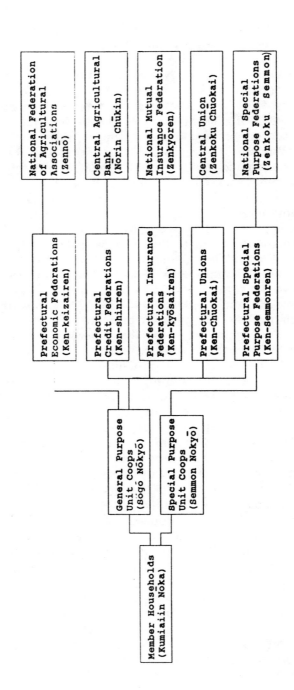

Figure 8.1
The Nōkyō Organization

AGRICULTURAL COOPERATIVES IN NAKADA TOWNSHIP

Unlike the village administrations that were merged into Nakada Township in 1956, the Nakada cooperatives have retained traditional village identities. Although the township administration and the respective Nōkyō worked out a merger in 1988, the fact that the four Nakada Nōkyō remained separate until that time was probably due to the historical hierarchy regarding water availability. The upstream Nōkyō had their roots as special purpose cooperatives while at least one downstream Nōkyō was started as a financial cooperative. The Nōkyō districts are shown on Figure 8.2.

The four local Nōkyō represent local land and water rights, responsibilities, and problems associated with agricultural production. For instance, in the Uwanuma Nōkyō area, the first cooperatives were formed in 1908 to coordinate cutting horse fodder out of the grasslands (genya) in the nearby marsh. Further, after the reclamation in 1908, the Uwanuma Cooperative was organized to buy supplies and improve soil quality in the reclaimed marsh area. The main problem being addressed by the cooperative was poor drainage which hindered spring planting.

From the 1920s when the Uwanuma Nōkyō was formed, and continuing until the late 1950s, the basis for cooperative formation was special purpose farming needs. Therefore, separate cooperatives evolved due to requirements for seed, manure, and new soil. For instance, manure was in short supply in the 1920s and the cooperative functioned to equally divide the existing supplies. Another example of an Uwanuma special purpose cooperative was the soil cooperative started in the early 1950s after the great flood. Noticing that yields rose the year after the flood, farmers reasoned that silt carried by the flood waters must have increased soil fertility. Therefore, they formed the Sakuraba Soil Cooperative (Kyakudo Kumiai) which used horses and carts to carry silt to the fields.

In the downstream area, the Asamizu Nōkyō was formed to confront financial problems caused by successive poor harvests resulting from lack of irrigation water. The egalitarian nature of the Asamizu Agricultural Cooperative is in large part due to the fact that tenancy was widespread in the prewar era and that pan-hamlet work teams were more common than in the upstream hamlets. Such agricultural work teams were formed to labor collectively on the paddies of absentee landlords. A few of these continued into the 1970s before the advent of the rototiller and rice transplanter made them impractical.

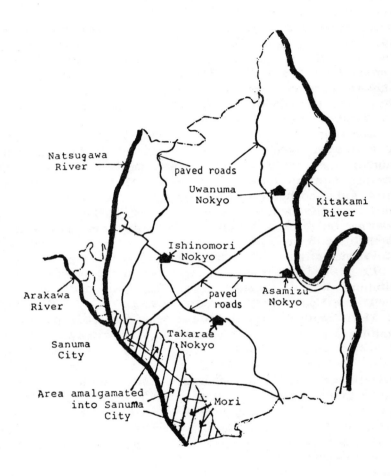

Source: Adapted from Miyagi-ken, Tome-gun, Nakada-chō 1980:109.

Figure 8.2
The Nakada Township Agricultural Cooperatives

The Asamizu Agricultural Cooperative was originally started by the Mizukoshi Trust Cooperative Society for the Utilization of Purchasing and Selling (Mizukoshi Shinyō Kōbai Hanbai Riyō Kumiai). At its inception in 1912 it included four hamlets with fifty-six members. It is significant that this cooperative originated within four years after the start of the Nakada Marsh Reclamation Project. As Nakada Marsh could no longer serve as a reservoir, the downstream inhabitants negotiated to maintain irrigation rights with the upstream Nakada Marsh Cooperative, which was controlled by upstream landlords. Historically there were many arguments over water rights and the downstream areas by necessity took turns using water cooperatively.

In 1912 the biggest problem for the Mizukoshi cooperative was borrowing capital for cash crops. It had few capital resources due to the drop in agricultural prices, natural calamities, and lower rice yields. As a result it was not able to borrow from commercial banks, and was compelled to collect money from the members each year. With this money it bought silkworms, fertilizer, and made improvements in mulberry orchards over the next three years.

Financing remained Mizukoshi cooperative's biggest problem. By 1926 the amount of interest on outstanding loans to financial institutions was 65 percent of the entire cooperative budget (Miyagi-ken Nōkyō Keiei Kenkyūkai 1966: 75).

Mori, another downstream irrigation area, is presently part of the Takarae Agricultural Cooperative. Because it was one of the lowest areas in the township, flooding was particularly severe. In fact, many of the older people can recall stories of marriages that were arranged with the hill hamlets of Asamizu because these hamlets would be able to send boats to rescue relatives during floods.

Mori contained an area of paddy land of which ownership was held in common. Because the sub-parcels differed on this land with respect to soil quality, drainage, and productivity, the farmers of Mori established a lottery system (kujibiki seido) whereby each year farmers were assigned a different plot. While this system was changed in 1960, it serves to demonstrate the collective approach to problem solving used in the downstream areas of the irrigation line.

The strength of social ties that result from villages that are in the same irrigation system is demonstrated by the case of Mori Village. In 1955 when an amalgamation law tried to force Mori Village to fall under the jurisdiction of Sanuma City, Mori and Nakada Township opposed. After two years of resistance, Mori was forced by prefectural decree to separate from Nakada Township and be incorporated into the nearby city of Sanuma. The farmers of Mori successfully resisted joining the Hasama Township (Sanuma) Agricultural Cooperative, giving as their reason that they were traditionally affiliated with the irrigation system of Nakada Township.

Currently, Uwanuma and Asamizu (Mizukoshi) Nōkyō have maintained separate identities. According to the Nōkyō annual reports of 1983, the Uwanuma Nōkyō had a much higher outstanding loan to deposit ratio than did the Asamizu Nōkyō, which was closer to the prefectural average. Uwanuma Nōkyō also had a low rate of member participation in the Agricultural Crop Insurance Program while the Asamizu Agricultural Cooperative had a high rate. The reason for this latter discrepancy is that the Uwanuma Nōkyō and the Agricultural Crop Insurance Program disagree over whether or not the flood zone adjacent to the river should be covered by insurance. Another dissimilarity between the Uwanuma and Asamizu cooperatives is the Nōkyō directors (riji) election. In the Uwanuma, Takarae, and Ishinomori Nōkyō there are fewer positions than hamlets to represent which results in a fierce election pitting hamlet against hamlet, while in Asamizu there is one director position for each hamlet.

Other differences between Uwanuma and Asamizu Nōkyō include sources of funding from various lending institutions (although both use the Central Cooperative Bank for rice) and attitudes concerning the Land Improvement Project. There was open opposition by Uwanuma Nōkyō youth leaders to the Land Improvement Project, while the Asamizu Nōkyō youth leaders were in favor of participation in the Land Improvement Project described in Chapter 10.

In 1970 Ishinomori and Uwanuma Cooperatives started a contract with three helicopters to spray 496 hectares of paddies for rice blast disease (imochibyō). Prior to helicopter spraying a farmer would have to team up with other farmers to spray an area. If the owner of the adjacent paddy did not control the disease, then the disease was likely to spread. This method of spraying was very effective for the control of rice blast disease and soon was expanded to the control of other diseases. The method was both cost effective and labor reducing although there was concern for the drifting of the chemicals into residential areas even though the spraying occurred in the early morning and people were alerted to stay indoors.

The two Nōkyō in the southern part of the township, Asamizu and Takarae Cooperatives, were proponents of a larger scale of farming which went hand-in-hand with the fact that these areas were more affected by the land reform and were the first areas to start the Land Improvement Project. Despite these differences, plans were underway in 1988 for the amalgamation of the four cooperatives in Nakada Township with the expected result of significantly lower management costs and improved services.

NŌKYŌ'S POWER BASE: RICE AND MONEY

Nōkyō controls rural farming finances because rice must be marketed through it. The 1947 Agricultural Cooperative Act changed the amount of rice that is routed through Nōkyō from 25 percent of the total registered rice in the prewar period (1937 figure) to approximately 96 percent in 1986 (Kahoku Shimpōsha 1981:57; Sankei Shinbun Henshūkyoku 1987:287). Nōkyō profits by charging rice marketing, inspection, and warehousing fees and oversees the flow of money from the government to the member producer.

There are three types of rice: government controlled rice (seifumai), independently routed rice (jishuryutsūmai), and free-market rice (jiyūmai). The government regulates rice by providing subsidies for seifumai and jishuryutsūmai. Figure 8.3 shows how rice is routed through the Nōkyō structure. In the event that free-market rice is unregistered, it is illegal and therefore called "black market rice" (yamimai).

The first type, seifumai, has its origins in the 1942 Food Control Act. Under this act, the government bought and sold rice in order to subsidize Japanese rice against the cheaper Taiwanese colonial rice which was flooding the domestic markets.

Independently routed rice, jishuryutsūmai, represents most of the remainder of Japanese rice produced. The subsidies for independently routed rice are, however, higher than government controlled rice because the price is dependent on the rice variety. The top varieties which are in highest demand and for which a shortage exists are Sasanishiki from Miyagi and Yamagata Prefectures and Koshihikari from Niigata Prefecture.

150

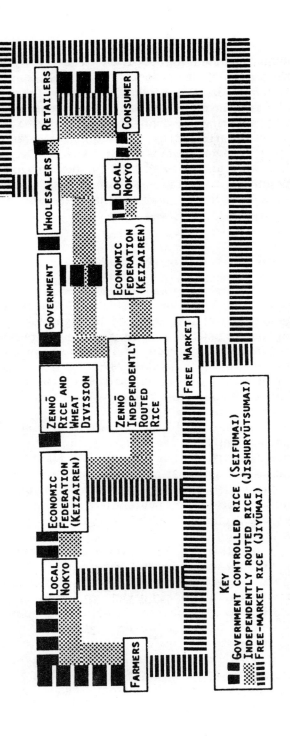

Sources: Kahoku Shimpōsha 1981, Tachibana 1980, Kome Mondai Kenkyūkai 1981,

Figure 8.3
Rice Routing in Japan

Government controlled seifumai rice is ranked according to grade with the top grade of government rice bringing a price several thousand yen less per sixty kilogram bag than the top varieties of independently routed rice. Farmers are, however, required to sell approximately half of their rice as government rice. The rest can be marketed as jishuryutsumai up to a certain limit (gendosuryo), which is defined according to productive capacity of the land. In 1983 in Nakada Township the gendosuryo for each ten are (0.247 acre) paddy was approximately eight bags weighing sixty kilograms each of hulled rice. In order for the government to lower its burden incurred through the rice price subsidies, the amount of seifumai has been gradually decreased. For example, in 1975 government rice constituted 86 percent of the total gendosuryo but by 1985 it had been reduced to 53 percent (Nihon Nogyo Nenkan Kankokai 1988:48,483).

Any remaining rice over the limit, can be sold as "over-the limit rice" (chokamai) and does not receive subsidies. While it was true that a severe rice glut reached its peak in 1982, since that time the surplus has been greatly reduced. In fact, as noted above, there is a shortage of high grade rice such as the Sasanishiki and Koshihikari varieties. Farmers growing these varieties or others can sell their extra rice through the free market for prices slightly lower than the government price. One such network is centrally located in the Kanda District of Tokyo (Kahoku Shimposha 1981:18). The availability of unregistered (black market yamimai) rice makes it possible for rice dealers to "blend" several grades of rice and still sell it as the higher variety. This mixing results in poor tasting "high quality" rice. This is why almost everyone in urban Japan puts such a high value on gifts of newly harvested (unblended) shinmai high grade rice from rural relatives.

The reason why Nokyo has evolved into a monolithic financial center has been its near monopolistic control over rice trading. In order to sell their rice, farmers must contract with Nokyo in April in order to receive from the government a down payment maewatashikin (also called gaitokin) to their Nokyo account in July. This amount equals approximately one-sixth of the price of each bag contracted. The remainder of the money is given to the farmer in December or January following the harvest. In this way farmers receive a sum from the government at roughly the same time that company employees receive bonuses. A number of farmers complain that the government should pay them sooner since modern mechanized farmers are now able to complete the harvest much quicker than they did twenty or thirty years ago and normally deliver their rice to Nōkyō during October. Farmers who are in need of cash sometimes try to sell rice on the free market because they can get their cash payments quicker.

On behalf of cooperative members, Nōkyō deals directly with the government Bureau of Food Supply, which administers the government buying and selling of rice. In Nakada Township during 1983, about 40 percent of the Uwanuma Nōkyō rice was sold to the National Grain Warehouse in the Edogawa District of Tokyo. The remaining 60 percent was sold to Chiba Prefecture, Shizuoka Prefecture, Nagoya, Ōsaka, Fukuoka, Kanagawa and Sapporo.

Nōkyō provides banking services and a line of credit to the farmers when they contract with the government for the seifumai rice. The farmers receive a down payment that is automatically deposited into a Nōkyō savings account. This down payment is paid by the government via the Agricultural Central Bank and prefectural Nōkyō Credit Federation to the local cooperative bank. Between April when they contract and December or January when they are paid, farmers can draw against their balance, and even overdraw if necessary before receiving the balance due from the government.

Because they have an automatic line of credit at 0.5 percent per month interest, many farmers prefer to buy machinery, fertilizers, chemicals, and groceries from the local Nōkyō store. Nōkyō offers credit several percentage points below independent dealers. Likewise, it is easier to qualify for loans at Nōkyō than through commercial banks because the farmer's rice crop and even farm land can easily be used as security for the loan.

The fact that farmers have a Nōkyō banking account by necessity draws them into the Nōkyō network of services. Table 8.3 shows the sale of items sold through the Nōkyō stores. Feed, foodstuffs, petroleum products, and fertilizers are the leading sale items.

Nōkyō figures prominently with the leading Japanese savings institutions. Farmers typically save about twice as much in Nōkyō as they do in the city banks and five times more than they do in the post office. Table 8.4 lists the amount of personal savings in various institutions in Japan showing that Nōkyō ranks fourth. Nōkyō's high ranking is a testimony to its dominance in rural areas. In lending out money to individuals,

Table 8.3

Sale and Handling Charges of Items Sold through Nōkyō Stores

Item	1984 Sales Amount (in million yen)	Nōkyō Handling Charges (%)		
		Nōkyō Unit	Prefectural Federation	Zennō
Feed	811,793	4.4	2.5	0.6
Fertilizer	498,195	11.0	2.3	0.6
Pesticides and herbicides	312,713	9.9	3.6	1.5
Agricultural machinery	363,707	9.5	4.2	1.5
Petroleum products	637,422	11.5	2.3	0.8
Packaging materials	138,423	8.3	3.0	1.3
Temperature control materials	104,129	8.5	3.0	1.3
Cars and trucks	221,631	6.1	1.9	0.7
Building materials	55,221	5.2	2.4	1.0
Rice	201,813	11.5	4.0	n.a.
Foodstuff (non-rice)	878,278	14.4	n.a.	1.1
Clothing	72,518	14.3	4.3	2.0
Durable consumer goods	109,628	11.9	4.7	1.7
Propane gas	137,406	42.4	6.4	0.8

Sources: Nihon Nōgyō Nenkan Kankōkai 1987:444; Sankei Shinbun Henshūkyoku 1987:287.

however, Nōkyō is second only to the city banks as shown in Table 8.5.

Farmers are drawn to Nōkyō for equipment loans to modernize their operations as well as loans necessary to pay for the Land Improvement Project described in Chapter 10. While about half of all agricultural loans originate from Nōkyō, if the government loans administered through it are counted, the figure is closer to 70 percent of the total.

The Nōkyō structure is similar to industrial groups (keiretsu) in Japan that are characterized by a number of companies each tied to a trading company and major bank. In the case of Nōkyō, Zennō serves as the trading company and Nōrinchūkin as the bank. In fact, like other Japanese banks, Nōrinchūkin has assets in the United States. As of June 30, 1987 it ranked sixteenth among Japanese banks in the amount of assets in the United States (Olstrom 1988:13). Nōrinchūkin draws much of its capital from the savings of local Nōkyō members which are in turn invested in the prefectural trust federations. These prefectural trust federations invest their money in Nōrinchūkin and securities. Although Nōrinchūkin ranks sixth among Japanese banks in total assets, it ranks first in domestic (yen) assets. Worried that farmers would take their deposits to higher yielding banks, in 1979 Nōrinchūkin began investing in high-yielding low-risk foreign securities. United States Treasury instruments account for about half of these. Nōrinchūkin supplies about one-fifth the total funds in the domestic call and discount markets which supply Japan's commercial banks and is one of the largest buyers of the Japanese national debt.

Tables 8.6 and 8.7 show the total sales and purchases of the three levels of Nōkyō. These levels are the local cooperatives, the prefectural federations (Keizairen), and the national level (Zennō). Keizairen and Zennō are used extensively in both selling and buying but are used slightly more in selling than in buying. It is also important to remember that although rice constitutes only 39 percent of the Nōkyō marketing total, the impact is much greater owing to the subsidies associated with it. Similarly it must be kept in mind that vegetables receive subsidies because in many cases they are grown as rice diversion crops (see Appendix A). Zennō is active in foreign wheat and feed buying, with granaries located on the Mississippi River.

Another source of Nōkyō's financial might is Nōkyō insurance (kyōsai). Although the amount of profit is about one-half that of Nōkyō banking, the profitability has been greater.

Table 8.4
Personal Savings in Savings Institutions in Japan in 1985

Institution Type	Amount (100 million yen)
Nōkyō	380,779
City Banks	501,200
Regional Banks	493,718
Mutual Banks	229,194
Trust Depository Banks	370,815
Post Office	1,029,978

Source: Zenkoku Nōgyō Kyōdō Kumiai Chūōkai 1987a:267.

Table 8.5
Personal Lending from Savings Institutions in Japan in 1985

Institution Type	Amount (100 million yen)
Nōkyō	107,150
City Banks	119,830
Regional Banks	84,923
Mutual Banks	53,190
Trust Depository Banks	75,825

Source: Zenkoku Nōgyō Kyōdō Kumiai Chūōkai 1987a:267.

Table 8.6
Nōkyō Sales in 1985 (in billion yen)

Item	General Purpose Nōkyō	Prefectural Federation	Zennō
Rice	25,894	25,843	25,634
Wheat	1,946	1,851	1,907
Vegetables	10,239	7,895	6,241
Fruit	7,144	3,657	2,626
Livestock	14,813	11,477	7,876
(Milk)	3,811	3,655	1,432
(Beef and Pork)	7,172	5,079	3,387
(Eggs)	1.093	1,158	1,795

Source: Nihon Nōgyō Nenkan Kankōkai 1988:336.

156

Table 8.7
Nōkyō Purchasing in 1985 (in billion yen)

Item	General Purpose Nōkyō	Prefectural Federation	Zennō
Fertilizer	4,896	4,270	3,838
Feed	7,468	6,363	8,920
Agricultural Machinery	3,784	2,584	1,595
Pesticides and Herbicides	3,292	2,178	2,397
Daily Necessities	18,553	14,325	3,501
(Food)	11,164	10,171	1,372
(Clothing)	733	744	520

Source: Nihon Nōgyō Nenkan Kankōkai 1988:337.

In fact, the National Federation of Agricultural Insurance Associations (Zenkyōren) has continued to lead non-agricultural insurance firms such as top-ranking Japan Life Insurance Company. As a cooperative, the profits can lower premiums and increase the dividends. Two types of insurance, long-term and short-term, are provided in areas including crop, life, house and building, retirement, fire and calamity, and automobile insurance.

RICE POLITICS

As one might expect, most of Nōkyō's political concern is focused on the rice price. By raising the rice price nearly all members benefit since most Japanese farmers raise rice and, the higher the rice price, the more money Nōkyō has circulating through its banking function.

Each year Nōkyō sponsors the rice price rally (beika undō) starting with each Nōkyō unit parading in its own district and then sending representatives to the prefectural and national rallies. Signs are posted throughout the countryside pushing for higher rice prices and opposition to the intrusion of foreign agricultural goods into Japanese markets. The signs are usually designed by the Women's Division and Young Men's Division of each local Nōkyō unit. The signs often reflect local humor, national pride and patriotism. Billboards in Miyagi Prefecture in

1983 included the following: "Don't Use Japanese Agriculture As Collateral For Cars," "If You Are Japanese, You Eat Rice", "Rice Is The Life Of Japan," "Even If There Is Too Much Rice, There Isn't Enough Sasanishiki Variety," "America Is Using Food As A Weapon To Land On Shore To Japan--The Country's Greatest Weapon Is The Farmers." In opposition to the beef imports one humorous sign read "Enough of American meat!" (Amerikasan no Gyūniku wa mō Takusan!) using the term "mō" (pronounced "moh") which has the dual meaning of "too much" and "moo" the sound cows make. Photo 8.1 shows the rice price rally and 8.2 depicts one of the rice price rally signs.

The rice price rallies were particularly effective during the 1960s when the rice price was rapidly rising but have become less effective in recent years as pressure has mounted on the government to stop subsidizing a commodity in oversupply. In 1985 the amount of stored rice was 1,030,000 metric tons, the storage and maintenance of which cost the government 4,561 billion yen, or 13.4 percent of the national budget for agriculture. The difference in the government buying and selling price of rice is shown in Table 8.8. Rice acreage is closely regulated as shown in the tensaku rice diversion policy described in Appendix A and Table 8.9. As a result of the rice gluts of the late 1960s and late 1970s, the amount of rice acreage was closely monitored and regulated by the government. The resulting diversion program costs and the cost of the rice price subsidies are shown in Table 8.10.

As can be seen from Table 8.8, in recent years the consumer's rice price has been rising at a rate higher than the producer's rice price. While this has helped maintain the high rice price for the farmers, it has fueled consumer resentment of farmers who they believe may have it too easy.

The rice price rallies are aimed at influencing the Rice Price and Farm Policy Deliberation Councils which have members from business, academic, and consumer groups. Each summer when it is time for deciding the rice price, Nōkyō presents the Rice Price Deliberation Committee with its draft of what the new price should be and the rationale for that price. It also tries to put pressure on the Comprehensive Farm Policy Research Committee of the Liberal Democratic Party (LDP). Both houses of the Diet have more than two hundred members on their agricultural caucuses.

Another reason why the lobbying efforts of Nōkyō are so effective is that rural districts are over-represented in the Diet

Photo 8.1
The Rice Price Rally: "Rice is the Life of Japan"

Photo 8.2
"Don't Sacrifice Agriculture for Industry"

Table 8.8
Changes in the Government Buying and Selling Price of Rice
1965-1985 (in yen)

Year	Producer's Rice Price	Government's Rice Price (A)	Consumer's Rice Price (B)	Cost Differential[a] (A-B)
1955	4,064	4,026	765	230
1956	4,028	4,024	790	266
1957	4,129	4,360	850	539
1958	4,129	4,356	850	539
1959	4,133	4,356	850	535
1960	4,162	4,351	850	506
1961	4,421	4,326	850	247
1962	4,866	4,877	955	365
1963	5,268	4,819	955	- 37
1964	5,985	4,783	955	- 754
1965	6,538	5,570	1,110	- 458
1966	7,140	6,063	1,215	- 521
1967	7,797	6,990	1,395	- 195
1968	8,256	7,551	1,510	- 39
1969	8,256	7,497	1,510	- 39
1970	8,272	7,442	1,510	- 55
1971	8,522	7,377	1,510	- 305
1972	8,954	7,846	1,590	- 190
1973	10,301	7,806	1,590	-1,523
1974	13,615	10,256	2,100	-2,001
1975	15,570	12,205	2,495	-1,771
1976	16,572	13,451	2,740	-1,379
1977	17,232	14,771	3,000	- 568
1978	17,251	15,391	3,125	157
1979	17,279	15,891	3,235	734
1980	17,674	15,891	3,235	339
1981	17,756	16,391	3,350	881
1982	17,951	17,033	3,482	1,446
1983	18,266	17,673	3,628	1,921
1984	18,266	18,327	3,764	2,256
1985	18,688	18,598	3,867	2,604

Note: The rice price figures are given for sixty kilograms of brown rice except in the case of the consumer's rice price, which is based on the price of ten kilograms of polished rice.

Source: Kanō 1987:218.

Table 8.9
Changes in the Rice Supply

Year	Rice Acreage (X 1000 ha.)		Production Index (Percent)	Domestic Production (X 10,000 tons)	Demand (X 10,000 tons)	Surplus Rice (X 10,000 tons)	Imported Rice (X 10,000 tons)
	In Production	Diverted					
1955	3,222		118	1,239	1,128	76	129
1956	3,243		104	1,090	1,166	36	56
1957	3,239		107	1,146	1,235	22	43
1958	3,253		108	1,199	1,217	28	40
1959	3,288		109	1,250	1,234	44	25
1960	3,308		108	1,286	1,262	50	22
1961	3,301		102	1,242	1,306	10	8
1962	3,285		105	1,301	1,332	2	18
1963	3,272		101	1,281	1,341	1	24
1964	3,260		99	1,258	1,336	5	50
1965	3,255		97	1,241	1,299	21	105
1966	3,254		99	1,275	1,250	64	68
1967	3,263		112	1,445	1,248	298	36
1968	3,280		109	1,445	1,225	553	27
1969	3,274	(5)	102	1,400	1,197	720	7
1970	2,923	(337)	103	1,269	1,195	589	2
1971	2,693	(541)	93	1,089	1,186	307	1
1972	2,640	(566)	103	1,189	1,195	148	0
1973	2,620	(562)	106	1,214	1,208	62	4
1974	2,724	(313)	102	1,229	1,203	114	6

Table 8.9 (continued)

1973	2,620	(562)	106	1,214	1,208	62	4
1974	2,724	(313)	102	1,229	1,203	114	6
1975	2,764	(264)	107	1,317	1,196	264	3
1976	2,779	(195)	94	1,177	1,182	367	2
1977	2,757	(212)	105	1,310	1,148	572	7
1978	2,548	(438)	108	1,259	1,136	650	5
1979	2,497	(472)	103	1,196	1,122	666	2
1980	2,377	(585)	87	975	1,121	439	3
1981	2,278	(668)	96	1,026	1,113	268	7
1982	2,257	(672)	96	1,027	1,099	90	6
1983	2,273	(639)	96	1,037	1,098	13	2
1984	2,315	(620)	108	1,188	1,094	32	17
1985	2,342	(594)	104	1,166	1,085	103	2
1986	2,303	(618)	105	1,165	----	--	--

Source: Kanō 1987:210.

Table 8.10
Government Subsidies Relating to Control of the Rice Supply (in 100 million yen)

Year	Ministry of Agriculture Budget (A)	Food Supply Control Program Costs (B)	Rice Crop Diversion Program Costs (C)	Food Supply Control Costs (D)=(B + C)	Food Supply Control Costs As a Percentage of the Ministry of Agriculture Budget (D/A)
1960	1,669	290	--	290	17.4%
1965	4,049	1,205	--	1,205	29.8
1970	9,921	3,746	818	4,564	46.0
1975	22,892	8,114	1,061	9,175	40.1
1980	37,765	6,521	3,034	9,555	25.3
1981	38,241	6,519	3,622	10,141	26.5
1982	38,207	6,402	3,652	10,054	26.3
1983	36,852	5,725	3,447	9,172	24.9
1984	34,844	5,403	2,683	8,086	23.2
1985	33,895	4,561	2,391	6,952	20.5

Source: Kanō 1987:216.

and are still very pro-LDP. Voting districts are weighted heavily in favor of the rural population making an added incentive for politicians to cooperate with the farmers. On a per capita basis a Diet representative from Miyagi Prefecture represents only one-half the number of people as a Tokyo Diet representative. In 1980 the rural prefectures excluding the nine most populous prefectures elected 60 percent of the lower house seats. Of the total LDP members in the lower house, these rural prefectures elected 69 percent. In the worst under-represented cases such as the Fourth District of Kanagawa Prefecture, there are three times more people per representative than in six of the most over-represented districts of Nagano Prefecture (Yomiuri Shimbun 1989:1). Hence, the political impact of Nōkyō's lobbying efforts reach well beyond the population of the farming households they represent.

Starting at the hamlet level, the rice price rally requires mandatory attendance of members rather than voluntary participation. This is due to the goal of the rice price rally which is to apply pressure on local politicians by ensuring representation by each hamlet at the prefectural demonstrations and by each township at the national demonstrations. The implicit threat to the politicians is that the hamlet and township group being represented will vote against them if they do not support the Nōkyō policy. When the farmers visit the prefectural seat or go to Tokyo to demonstrate they also visit local representatives to make sure they will support efforts to increase rice subsidies.

In order to subsidize the rice price rallies Nōkyō charges it's members according to how much they use Nōkyō for marketing. For instance, the Uwanuma Nōkyō charged thirty yen per bag of rice. This provided funds for sending representatives to Sendai and Tokyo. Uwanuma Nōkyō allocated sixty-five thousand yen for two members to attend the Tokyo rally. About four times that amount was spent for fifty-two members to attend the prefectural rice price rally in Sendai.

Nōkyō evaluates each candidate's willingness to cooperate by whether the candidate is sitting up front on the platform at the rice price rally and sponsoring and voting for Nōkyō favored policies. In Miyagi Prefecture, eight of the eleven Liberal Democratic Party diet members in 1979 were self-proclaimed "pro-farming" members. As shown in Table 8.11, these members were evaluated on the basis of their committee support and their actual attendance at the pro-farm rallies.

Table 8.11
Miyagi Prefectural LDP Diet Members and Farm Policy Support in
1979

	Committee Membership	Nōkyō Rally Attendance	Total Number
Lower House LDP Reps.			
Kazuo Aichi	1	1	2
Takashi Hasegawa*	1	3	4
Shūichiro Itō	5	4.5	9.5
Fukujiro Kikuchi	2.5	1	3.5
Hiroshi Mizuka	3	2	5
Hideo Utsumi	5.5	1.5	7
Upper LDP House Reps.			
Kaname Endō	1	3.5	4.5
Buichi Ōishi*	4.5	4	8.5

*Refers to the major politicians who affect K Hamlet politics.

Source: Tachibana 1980:340.

In 1979 Nōkyō initiated the rice diversion program to grow various crops on land formerly used to grow rice (see Chapter 11). This was an attempt for Nōkyō to gain status as an equal partner in negotiating with the government on agricultural issues. The government approved the plan with some modifications but has not allowed Nōkyō to enter other areas of agricultural decision-making.

On the Nakada Township level it can be said that the head of Nōkyō possesses nearly equal status as the mayor. In Nakada Township, agriculture revolves around a triangle formed by the three administrative units which control government policy, water and land improvement, and buying and marketing functions. These are the township office (yakuba), the land improvement office (kairyōku), and the agricultural cooperative (Nōkyō). The land improvement district is more autonomous than either the town office or Nōkyō and independently solicits project grants. In Nakada Township the land improvement office works closely with the township office particularly with respect to the Land Improvement Project.

It is significant that whenever a prominent politician such as a cabinet minister passed through Nakada Township, both the Nōkyō head (rijichō) and the mayor of the town gave speeches in his honor. The riijichō, however, spends the greatest share of his time planning and goal setting. In January he meets with each of the sub-hamlet groups when they make their over-night trip to a local spa. In July he is busy negotiating with politicians concerning support of the rice price. In addition, he must be a representative to the prefectural Central Committee in Sendai where he is very active in rice politics, which includes making agreements concerning the routing of rice, financial arrangements with other banking institutions, and negotiating with politicians for project money (see Appendix A for examples of how many of the crop diversion program projects can be funded through Nōkyō). He also can often be seen at the Land Commission office and the township office.

The Nōkyō head is elected from among the Nōkyō directors who in turn are elected through a fierce competition for votes in a local election. These directors (riji) have almost an equal status with the hamlet head (kuchō or burakuchō), another demonstration of how important Nōkyō can be in the rice belt of rural Japan. Both riji and kuchō in K Hamlet were expected to attend weddings and give a bottle of sake to the senior citizens association when they made recreational trips to the spa.

In Nakada Township each of the cooperatives has a slightly different electoral process. As might be expected owing to its cooperative past, in Asamizu the number of directors equals the number of hamlets. However, in the Uwanuma Nōkyō there were only twelve director positions with sixteen participating hamlets.

Thus, in the Uwanuma Nōkyō election, hamlets tried various strategies to capture votes. These methods included making deals with individuals from other hamlets individually or en masse. Sometimes vote buying was used at a rate of approximately twenty thousand yen per vote. However, since this was expensive, an alternative method was to increase the number of voting members per household. K Hamlet, for example, had ninety-one full members with fifty-eight farming households after a campaign to register spouses during the period 1981 to 1983. Adding suspense to the final outcome of the vote, sometimes daughters who had married into other hamlets were telephoned to encourage them to cast their vote for their home hamlet. The Nōkyō election in which each hamlet casts its ballots in turn has a much higher voter turnout than the town council election or the prefectural and national elections.

NŌKYŌ AND GRASS ROOTS SOCIAL ACTIVISM

Notwithstanding complaints that Nōkyō had grown too large and impersonal and that costs were not as low as they should be, in Nakada Township most farmers view Nōkyō as a grass roots farming organization which rivals local town government and land improvement district offices in influence. The heads of the four Nōkyō share the podium with the mayor when important government dignitaries visit the township. The greatest symbol of grass roots identity is the annual Nōkyō sports day, which is more like a festival than a competition.

The sports day features fun games that emphasize traditional culture. For example, <u>dojo tsukami</u> requires the participant to run a short distance and then catch with one hand a loach swimming around in a bucket. (The farmers enjoy catching loaches in the paddy drainage ditches after a summer rain.) Another game, <u>nawanae kyōsō</u>, has two-person teams making straw rope which is then tied to a sack which must be pulled across the finish line. Men can try their luck at the rice bag lifting contest which requires the contestant to hold a traditional straw wrapped bag of rice over one's head as long as possible. Other races test various combinations of age and sex in both individual and relay races.

The most important thing about the sports day competition is the fact that each of the hamlets competes as a group against the other hamlets. Both farming and non-farming households participate and an effort is made to allow representatives from each household to take part. Each hamlet erects a tent next to the track which serves as a center from which the families cheer on their hamlet representatives in the competition. Nōkyō sports day activities are shown in Photos 8.3 and 8.4.

Midway through the competition, the Nōkyō officials give a short speech thanking the community for their participation. Following this is the half-time entertainment competition. Each hamlet does a dance or makes a float. In 1983, the hamlet that won the contest made a float in the form of a palanquin carrying a traditional rice bag with some rice plants that had been pulled up just before the contest. Playing on the theme of the local custom <u>ueage</u>, a rice transplanting rite in which some rice plants are uprooted and offered to the Shinto gods, the sign on top of the rice said "Prayer to a Plentiful Harvest." This was especially appreciated owing to the fact that the previous three years were poor harvests.

Photo 8.3
Winning Hamlet Float at the Nōkyō Festival. Sign on top of straw
bags of rice reads "Prayer for a Plentiful Harvest."

Photo 8.4
Nawanae Kyōsō (Race to Make Straw Rope). Representatives of a
main and branch household work together in the competition.

THE FUTURE OF NŌKYŌ

The future success of Nōkyō is dependent on whether or not it can maintain its central role in rural Japanese society and maintain an image of grass roots participation and cooperation. While Nōkyō has been active in promoting activities for agricultural youth, and addressing problems such as that of recruiting farm heirs and marriage partners described in the next chapter, the future of the organization will depend on how well it can adapt to the rapidly changing social, political, and economic environment.

Despite the fact that the youth organization of Nōkyō played a key role in organizing farmers in the 1989 Upper House election in which the LDP was soundly defeated by the Socialist Party, the problem of how well Nōkyō represents the interests of the farms population still lingers.

The organization has shifted away from total dependence on rice. In 1970 rice constituted 51.3 percent of its total earnings but had dropped down to 34.9 percent in 1981 (Nihon Nōgyō Nenkan Kankōkai 1983:421). Recently it also approved the rice diversion policy (tensaku) which it had previously opposed because tensaku stemmed from earlier policies of rice acreage curtailment (gentan). Increased handling of vegetables and livestock are filling the gap in rice earnings.

Nōkyō has tried to maintain the institution of small-scale farming in Japan and thereby keep its central role. It has had some degree of success in increasing the number of urban Nōkyō. If, however, rice is deregulated from under Nōkyō's near monopolistic control, or if foreign rice is imported, Nōkyō will have difficulties maintaining much of its operation.

Likewise, the city and trust banks in many cases are offering better services to customers. Nōkyō will need to keep pace with the modern banking world. Many Nōkyō are using the on-line computerized systems of city banks because they are more efficient than the currently used system.

Although Nōkyō can diversify and modernize, this in itself will not be enough. The key issue for Nōkyō will be how to maintain a stable farming society based on small-scale producers. The number of farmers will further diminish if farmers continue to face the snowballing debt cycle caused by the need to buy expensive modern equipment. Likewise, the problem of securing spouses for heirs is very real and if not resolved, will affect not only the membership in hamlet and sub-hamlet (jikkō kumiai) units but the continuity of the rural household inheritance system. Apathy and the "free-rider" problem may also grow as members see the cooperative not as a grass-roots organization protecting village interests and culture but as an amalgamated profit-seeking financial giant which tries to save management costs in order to maintain its structure.

NOTES

1. Takekazu Ogura (1982), architect of the Agricultural Basic Law, has a detailed review of the history of the ideas behind the development of agricultural policy in Japan.

2. This development was a significant one that followed on the footsteps of the tenant suffrage right. Prior to this only male landowners had the right to vote.

9

Industrial Development and the Rise of Rural Subcontracting

INTRODUCTION

The growing importance of industry in rural Japan today is revealed by the small number of farmers in the field on weekdays. Instead, farmers turned blue collar workers face morning rush hour traffic. In 1985 the average Japanese farming household earned only 15 percent of its income from farming (Nihon Nōgyō Nenkan Kankōkai 1987:536). By combining wage labor earnings with agricultural income, the average farming household in Miyagi Prefecture earned 6,928,000 yen ($55,424) per year, which was very close to the national average for farmers (Nihon Nōgyō Nenkan Kankōkai 1987:473).

The expansion of second-tier companies and subcontracting companies into rural areas provides opportunities for farming families to supplement their declining farm income and even surpass the level of income attained by their counterparts in the city. This pattern of decentralized industrial development prevented the rapid demise of the farming population which occurred in Europe and the United States, and in 1985 Japan still maintained over 16 percent of its population as farmers.

Some explanations of Japanese industrial success have emphasized low wages and a high degree of subcontracting. While I found this to be true in Miyagi Prefecture, this alone would not separate the Japanese case from that of Korea or Taiwan. The real key to Japanese rural industrial success is the remarkable social stability of households in the farming community. This stability has been reinforced through the practice of single heir inheritance. Industry in rural areas has

171

favored hiring heirs and their spouses as a recruitment strategy because heirs and their spouses have moral obligations to their ancestors regarding the maintenance of land rights and are therefore not likely to move out of the township. Such employees are expected to maintain company harmony because in many cases personal family connections are used to make the initial contact for the job interview. According to a study carried out by the Agricultural Land Commission (1981), 74 percent of the respondents were introduced to their job through either relatives, friends, or acquaintances. Because of the social responsibility towards these personal connections, company stability is enhanced.

This part-time farming phenomenon arose in three stages of industrial development, utilizing the labor that became available as a result of the time saved by mechanizing rice agriculture. The stages and their main characteristics are listed below:

Stage One: 1960-1972 Seasonal Labor Migration to Urban Areas
Stage Two: 1972-1984 Decentralized Industrial Development
Stage Three: 1985-present   Global Competition

This chapter argues that during the stages of seasonal migratory labor and decentralized industrial development, the increased availability of jobs and their similarity in pay was a factor that decreased economic differences between households. This was especially true in closing the economic gap between landed and landless households, a gap which had already narrowed due to the establishment of upper limits on land holdings and attempts to equalize land holdings between tenant and landlord households by the land reform.

The trend towards economic equality has been tempered by recent developments in the rise of the value of real estate, the yen appreciation, and the threat that the subcontracting plants may close. Minor differences in agricultural land scale became magnified due to the rapid rise of real estate prices between 1963 and 1974. The devaluation of the dollar between 1985 and 1987 forced local industry to cut production costs through automation at home and investment in foreign plants abroad. Suddenly, the job security offered through second-tier kogaisha "child" companies became economically more important as workers in subcontracting firms feared losing their jobs through automation.

Because industrial income constituted an overwhelming share of household income, both landed and landless households were concerned over job security in the manufacturing sector. Owing to the fact that there is a high correlation between landed households having workers in kogaisha and landless households placing workers in subcontracting companies, for the first time since the war, economic differences within rural communities became a possibility if not a reality. This trend disrupted economic and social equality and increased the difficulty for continuing the ie household system.

This chapter will outline the stages of industrial development in Nakada Township, including the gradual development of dekasegi "seasonal jobs" available in cities during the 1950s and 1960s, the rise of decentralized second-tier kogaisha and their subcontracting vendors in rural industries in the 1970s, and the present stage of international competition. Additional consideration will be given to the social problems posed by the threat of rapid automation and the kūdoka "hollowing out" of Japan.

STAGE ONE: SEASONAL LABOR MIGRATION FROM RURAL TO URBAN AREAS

Following World War II, there were few opportunities for seasonal work in Nakada Township, except in the local silkworm industry. As a result, most farming households continued to use mixed cropping (rice plus another crop) in order to fully utilize household labor. Still, the off-season for farming in Tōhoku was in the late fall following the rice growing season and preceding the spring planting. As shown in Figure 9.1, dekasegi seasonal employment opportunities became available in the 1960s and lasted until the early 1970s, at which time better paying jobs became available close to home.

In 1956 there were a total of only eighty-six dekasegi workers from Nakada Township traveling outside the prefecture for work. Seventy-four migrated outside the prefecture, while twelve found seasonal jobs within. By 1965, only nine years later, nearly a quarter of all farming households in Nakada Township had a family member migrate during the agricultural off-season. Almost all dekasegi workers were male. Of these, 47 percent migrated to Tokyo and the remainder went to Kanagawa Prefecture, Hokkaidō, Shiga Prefecture, the Tōkai Region, the Hokuriku Region, and other prefectures in the Tōhoku Region.

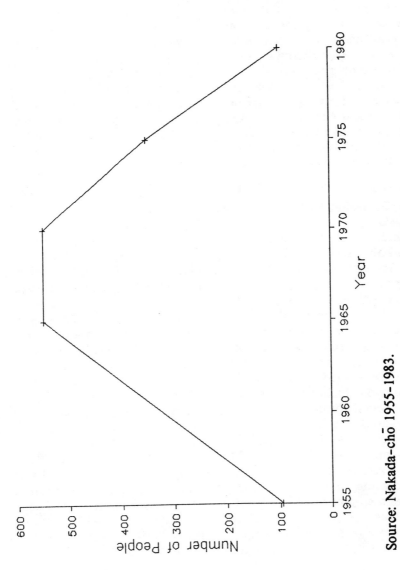

Source: Nakada-chō 1955–1983.

Figure 9.1
The Rise and Fall of Seasonal Migratory Labor from Nakada Township

Future research is needed to document the degree of contact between <u>dekasegi</u> workers and their urban relatives. This is important because the <u>ie</u> system forces non-heirs out of the household while remaining socially obligated to help these individuals in time of need. Likewise, these non-heirs maintained their ancestral ties to their <u>furusato</u> "rural homeland" by attending annual <u>Obon</u> ceremonies and anniversaries of funerals as well as helping out with spring planting (before the advent of the transplanting machine in the 1960s). This open line of communication facilitated seasonal migration back and forth.

The number of <u>dekasegi</u> workers reached its peak in 1971 shortly after the township established the Farmers Human Resource Placement Bank to help farmers find seasonal work and other non-farming occupations. However, this placement center was short-lived and the rate of <u>dekasegi</u> migration steadily declined thereafter.

STAGE TWO: DECENTRALIZED INDUSTRIAL DEVELOPMENT

During the early 1960s, the New Industrial City Promotion Law and the Special Regional Development Promotion Law were passed to decentralize industry away from the Tokyo-Osaka regions to the rural areas. The trend toward decentralization was later emphasized in 1972 by former Prime Minister Kakuei Tanaka in his plan for remodeling the Japanese archipelago and then in 1979 in <u>Sanzensō</u>, the Third Comprehensive National Development Plan. The idea of decentralized industrial development has continued into the 1980s with the Technopolis Concept developed by the Ministry of International Trade and Industry (MITI) and <u>Yonzensō</u>, the Fourth Comprehensive National Development Plan described in Chapter 12.

Leading Tome County in the national trend promoting rural industry, Nakada Township bought up a section of land in 1967 to offer to prospective industries. These industries were required to possess fixed capital of over 500,000 yen and employ over thirty permanent employees, or 200,000 yen fixed capital with over ten permanent employees. The first attempt at attracting a company (the Atsugi Nylon Factory) ended in failure because the underground water contained too much iron for the company's needs.

Critics of industrial development were quick to point out that the township had made a great financial mistake by acquiring the land. Mayor Hachirōzaemon Miura, however, continued to invite prospective industries to Nakada Township. By the end of his administration, which lasted from 1960 to 1971, Mayor Miura had managed to persuade six factories to locate in the township. He had also made arrangements for the establishment of two Sony factories, which located in the township the year after his third term.

Between 1966 and 1979 the number of persons employed in these industrial firms increased six-fold. After 1975, there was a slowing in the number of companies to locate in Nakada Township. Only six factories have arrived since then and all these hired fewer than one hundred employees. The list of the factories and their locations are given in Table 9.1.

Nōkō Ittai: The Unity of Agriculture and Industry

Under the progressive leadership of Mayor Miura, Nakada Township invited several scholars to the township in 1966 to lecture to the Nakada Township Construction Deliberation Council. The lecture series was entitled "How Should We Plan for the Shape of the Future for Nakada Township: In What Way Should We Plan for a Bright, Prosperous Township?"

One of the invited speakers, Hotarō Takeda, was the director of the Research Institute for New Agricultural Policy (Shinnōsei Kenkyūjo). He chose the topic, "The Consciousness Revolution of The Agricultural Villager--Toward the Construction of the Ideal Agricultural Village." In his presentation he listed three reasons which he felt called for "revolutionizing" agrarian villages. First, the agricultural population was rapidly declining and without some kind of change the township would lose its younger population to the cities. Second, foreign agricultural trade seemed inevitable because Japanese production costs were well above the international level and because there was growing consumer demand for foreign products. Last, the Japanese diet itself was becoming Westernized.

His proposal consisted of two aspects: modernization of agriculture and the promotion of rural industrial development to unify agriculture and industry (Nōkō no Ipponka). In his plan, modernization was defined as "increasing productive capability and lowering costs" (Kōhō Nakada 1967) and his goal to

Table 9.1
Industry in Nakada Township

Company Name	Total Number of Employees	Percent Female	Year Founded	Type of Industry
(Sony) Sound Magnetics, Inc.	435	83	1972	Electronics
(Sony) Nakada Magnetics, Inc.	354	65	1972	Electronics
Renown Look Nakada Plant	146	95	1968	Women's Apparel
Miyagi Production Plant of Stanley, Inc.	140	80	1970	Auto headlights, Camera light meters
Asahi Sewing	102	88	1970	Jeans
Nihon Electrical Appliances	62	94	1977	Camera shutters
Towa Acoustics Nakada Plant	57	n/a	1985	Video parts
Tehoku Chiyoda Miyagi Plant	57	14	1968	Electrical parts
Nagane Electronics	48	94	n/a	Cassette radio mounts
Ishinomori Electronics	45	93	1968	Car stereo mounts
Shinwa Electronics	45	n/a	1982	Electronics
Nakada Sewing	44	100	1979	Jeans
Nakada Electronics	42	88	1978	VTR assembly
Miyagi Limestone Manufacturing	37	21	1978	Limestone
Kawasaki Communications Nakada Manufacturing Plant	35	n/a	1983	Electronics

(Continued)

Table 9.1 (Continued)

Hasama Concrete Limited	35	9	1968	Concrete
Uwanuma Plant of Miyagi Carbonizing	34	21	1948	Limestone
Yokohama Processing	33	32	1968	Bathtubs and refrigerating units
Yamauchi Livestock	33	85	n/a	Meat products
Aoki Electronics	29	n/a	1985	Electronics
Tomei Textile Manufacturing	27	72	n/a	Womens' clothes
Nakada Fashions	26	95	n/a	Womens' clothes
Takahashi Electronics	22	n/a	1985	Electronics
Rodeo Sewing Nakada Plant No. 7	21	95	1973	Childrens' clothes
Asamizu Electronics	20	100	1973	Electronics
Nakada Broilers	19	n/a	1983	Broilers
Otori Manufacturing	17	53	1973	Stainless steel sinks
Chiba Sewing	16	77	1977	Skirts
Chiba Electronics	16	n/a	1986	Electronics
Ishinomori Trading Head Office	15	80	1973	Ski wear

Source: Nakada-chō Shōkōkai 1983, 1986; Nakada-chō Shōkōkai Mura Okoshi Jigyō Jikkō Iinkai 1986, Nakada-chō Shōkōkai, Nakada-chō Shōkōkai Fujinbu 1985.

modernize agriculture included decreasing the reliance of agriculture on national price subsidies. By increasing the use of large-scale machinery, agricultural productive capability was to be increased with a corresponding lowering of cost per unit produced. He also advocated deep plowing, crop rotation of rice and hay, better drainage facilities, and large amounts of fertilizer top dressing (four thousand kilograms per paddy).

As for rural industrial development, Takeda held that "industry would absolutely forge ahead into rural areas" (Kōhō Nakada 1967). The reason he gave for this was that cities gradually deny people a way to live by injuring their spirit and bodies, which he attributed to the over-centralization of industry in cities. He said that for the sake of the inhabitants as well as the industries themselves, industry would have to decentralize into the rural areas. He held that with a modernized agriculture and decentralized industry, a new type of civilization would emerge under the theme of unified agriculture and industry.

The following year Mayor Miura openly promoted the concept of the unity of agriculture and industry (nōkō ittai). New approaches toward agriculture and industry soon followed. Adhering to Takeda's goal of agricultural modernization, the Nakada Construction Deliberation Council conducted an experiment with large-scale machinery, the first such effort in Tome County. Approved in 1967, the experimental project was to take place between 1970 and 1974, and was designed to test large-scale equipment on 3.4 hectares of land in the traditional paddy environmental zone (kiseiden). If the experiment had proven successful, Nakada Township would have been a leader in large-scale farming.

The experiment speaks mainly for Miura's progressive administration and willingness to try new approaches. It showed that large equipment did save labor time but also revealed that the overall yield dropped due to mechanical errors. It also proved that the cost of the equipment would be higher than originally expected. Finally, the experiment was abandoned as a failure in 1974.

The Electronics Factories

By 1987 the two Sony factories together employed 789 persons which represented about 36 percent of those employed in Nakada Township. These factories, the second and third most heavily capitalized factories in Nakada Township, respectively, were branch companies established by a Sony factory near Sendai. Sony Saundo Magune (Sound Magnetics) made video tape recorder (VTR) and tape recorder heads. Sony Nakada Magune made ferrite for transformers used in household electrical goods (such as for televisions or video displays) and also manufactured floppy disks.

Judging from the number of employees and ownership, the two Sony factories and Renown Look can be classified as kogaisha, "child companies." They were entirely owned by their parent companies and heavily capitalized. The importance of a job in such a large company cannot be overemphasized. Nationally, workers in companies having fewer than ninety-nine employees received only 93 percent of the wages of employees in larger companies. Likewise, retirement benefits, on-the-job training, worker's compensation, and housing allowances were less in small companies. Most important, job security quickly diminished as the size of the company decreased.

The electronic and textile industries were characterized by a high degree of subcontracting. In Japan's manufacturing sector, more than 99.5 percent of the total number of firms had fewer than three hundred employees. For example, Toyota Auto Company had three thousand small and medium-sized subcontracting companies supply it (Kenmochi 1987:31). Brother, originally famous for its sewing machines, but now its computer products, had about two hundred subcontractors beneath it. The electronics industry in Japan had more levels in the multi-tier structure than does the automotive industry. Normally, a major electronics company had a number of kogaisha, each with subcontracting firms descending as many as four steps removed from the kogaisha. Textile companies had a simpler structure. Figure 9.2 is a flow chart illustrating a typical electronics subcontracting network.

In the Sony example, the parent company was sole owner of the stock in the child company and its two sibling companies. About 80 percent of the child company's employees were classified as regulars and the remaining 20 percent were classisfied as temporary. Female regular entry level line workers made about ninety thousand yen per month per six hour shift working four days on and four days off. Employees received three months bonus per year for total earnings of about ten thousand dollars per year. This kogaisha previously required its employees to work eight hour shifts but adopted six hour shifts

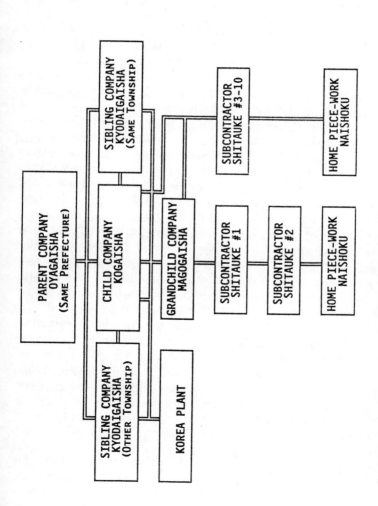

Source: Author's 1983 K Hamlet interviews.

Figure 9.2
Multi-tiered Subcontracting and Fictive Kinship in the Nakada Township Electronics Industry

to avoid laying people off in the face of the yen appreciation, which had reduced the company profits.

The amount of paid vacation increased with the number of years employed in the company. For one to two years of work, twelve days of paid vacation were awarded. Each year up to the tenth year one more day was awarded up to a maximum of twenty days. Under conditions of normal growth, regular employees working eight hours per day received between four and five months salary bonus per year and therefore made a total of about seventeen thousand dollars. In 1987 temporary male employees received no bonus and made about six thousand yen per seven hour day or $12,800 per year.

Shitauke Subcontracting Level and Home Piece-Work

The number of employees in subcontracting companies normally ranged from a low of twenty employees to about one hundred employees in the companies closest to the child company. These smaller companies utilized a high percentage of female workers. Usually for every two employees of a subcontracting company, there was one piece-work supplier working at home.

Female employees at the lower end of the electronics subcontracting network were paid 3,480 to 3,600 yen per eight hour day, which was very close to the prefectural minimum wage. They received no bonus so therefore earned about $7,680 per year. Males earned about 5,600 yen per day or $11,947 per year. Normally no paid vacation was given although sometimes pay incentives were awarded for perfect attendance during the month.

Home piece-work was paid according to output (number of widgets produced) and pay was usually less than that of the companies on the lowest rung of the subcontracting ladder. According to interviews, employees could solder enough widgets to earn somewhere in the three thousand yen per day range, or $6,400 per year. As a rule, the age of workers doing home piece-work was higher than that of other industrial employees. In K Hamlet, only four women and two men were engaged in home piece-work, although many families had participated in it at one time or another.

Reducing the Risk in Subcontracting

There was a further distinction between two kinds of electronics subcontracting companies, namely, ones which received their work orders directly from the child company and those that did free-lance work. The benefit of direct work was that it was reliable. However, the profit margin was somewhat less--hence the term "hosoku nagai" meaning "thin and long," referring to the thin profit margin and long duration of the work contract. Profits were "thin" owing to the fact that in these situations the parent company retains the right to determine pricing. Direct work also received quick bank payments from the child company. Normally, in this type of arrangement the subcontracting company was compensated in about two weeks. In free-lance work, although the owner was provided a better profit margin and more freedom to switch jobs, payment sometimes took up to thirty days, which in turn was in the form of a four month bank promissory note. This made the effective payment period about five months. In order to reduce risks, some subcontracting companies tried two ventures: one in something that was "thin and long" and one that was "short and fat."

Another method of reducing risk was relying on relatives. Many of the subcontracting companies were formed around deep-rooted social bonds between major household groups such as the dōzoku. These household ties were usually built around several marriage alliances. A case in point was the household alliance between two of the subcontracting companies supplying one of the Sony factories. The companies exchanged employees and work load according to supply and demand. They also called upon each other when recruiting employees. This form of hiring was a continuation of the past when they made alliances as leading landlords and silk worm producers.

Hiring Preferences

About 75 percent of all industrial jobs in Nakada Township were held by women, and electronics subcontracting firms employed the highest percentage of females. Two firms employed only females. While females have been able to find jobs in electronics and textiles, males have been employed more often by the government and by service industries, which have rapidly increased in number as shown in Figure 9.3. Both these

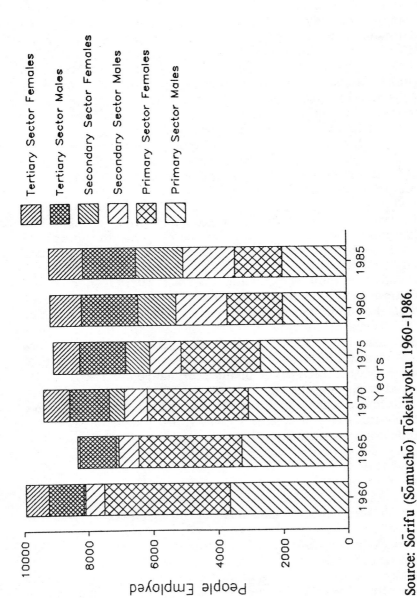

Source: Sōrifu (Sōmuchō) Tōkeikyoku 1960–1986.

Figure 9.3
Changes in Nakada Township Employment by Sector and Sex

occupational groups received slightly higher wages than did workers in the manufacturing industry.

All companies in Nakada Township gave special consideration to the needs of farming households and hired about 67 percent of their employees from within the township. Many provided microbuses to pick up and drop off employees from their homes. In addition, the Sony factories gave preference to eldest sons and their brides when hiring male and female employees. They also sponsored Obon dances and allowed employees to take paid leave for obligatory attendance at weddings, spring planting, and the fall harvest. Both large and small companies normally went on a company trip at least once a year.

STAGE THREE: GLOBAL COMPETITION AND THE IMPENDING CRISIS FOR RURAL INDUSTRY

In 1987 the gradual appreciation of the yen forced export oriented industries to either cut domestic production costs or produce in a foreign country. This situation severely affected electronics related companies such as Sony which export about 70 percent of its products. Figures 9.4 and 9.5 depict the yen appreciation and corresponding drop in profit for Nakada Sony factories. For the Sony company planners in Nakada Township, each Friday in 1987 was devoted to secret meetings where discussion of viable strategies for lowering production costs was made. According to interviews conducted in 1987 at the township office and at the chamber of commerce, the electronics industry in general was down about 25 percent. The textile industry seems to have been less affected (Nōsonchiiki Kōgyō Dōnyu Sokushin Sentā 1987).

Sony planned two strategies for lowering production costs. The first method of cost reduction (gōrika) was rapid automation (jidōka). Robots were to be introduced in large numbers in the next five years to replace much of the subcontracting work. Sony's competitor, the Toshiba Corporation, successfully reduced costs by 20 percent through automation in its Fukaya VCR plant (Chira 1988:25).

Owners of small electronic subcontracting firms challenged this idea saying that "There are just some things that machines can't do." One owner of a small electronics firm which made products sold under the Radio Shack Realistic brand name in the United States noted that already one-half of the parts found on

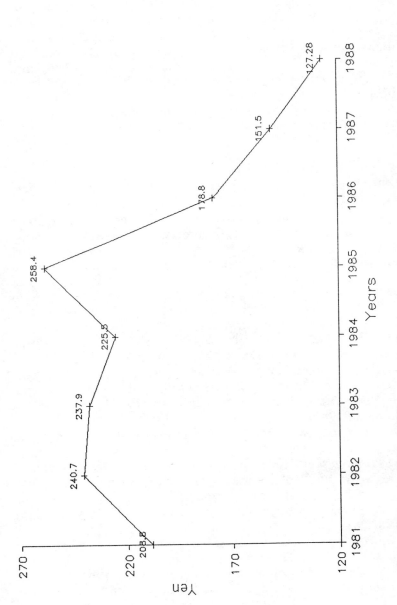

Source: Fuji Bank 1981–1989.

Figure 9.4
Yen Appreciation 1981–1988

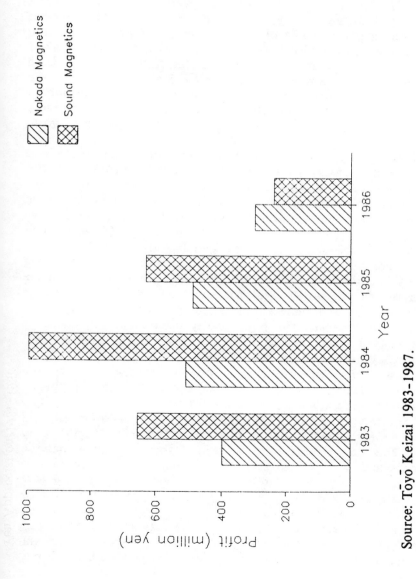

Source: Tōyō Keizai 1983–1987.

Figure 9.5
Profit Decline Among Nakada Sony Factories 1983–1986

the electronics boards that he received were filled in by robots. As an intermediate step some companies have required subcontracting companies to relocate inside the factory grounds, which facilitated "just-in-time" production and made it easier to automate.

A second cost-reducing technique was to build a plant in Korea. In Korea the 1987 labor costs were approximately one-fifth that of Japan. Sony, along with Sanyo and Hitachi, has also started a plant in one of Mexico's maquiladors, or industrial zones on the U.S.-Mexico border that take advantage of the $2.91 per day wage labor and the United States tariff code provision that allows duty-free movement of United States goods across the border provided that the items are re-imported in assembled products (Carlson 1987:21). In the future it is likely that more electronics work will be sent to Korea and the other third world nations.

This situation contrasts with the automobile industry where Japanese companies have relocated to the United States (principally the Midwest) to avoid voluntary auto import restrictions. Because of the emphasis in the United States on local part content, companies like Honda in Marysville, Ohio have been aiming for 75 percent local part content by 1990 and have even started reverse importation back to Japan.

In contrast with the American experience of moving assembly plants abroad, the Japanese, with their extensive subcontracting network, are focusing on moving the parts industry abroad and maintaining as much domestic assembly as possible in order to control high quality production and maintain job security for Japanese. The Bank of Japan reported that parts imports increased by more than 40 percent during the fiscal year 1986-1987 (Chira 1988:25). Sony has been a leader in the transnationalization process of the Japanese electronic appliance industry. In fact, during the period between 1986 and 1988 Sony was expected to double its overseas production to nearly 40 percent of its total production (Kitazawa 1987:17). In the Sony case in Nakada, it remains to be seen if this will mean that the Nakada plants continue to assemble the parts made abroad or if the assembly process will also eventually be exported. Figure 9.6 shows how the Nakada Sony industrial structure will be affected in the immediate future.

For the workers and owners of the Nakada subcontracting plants, kūdōka (the "hollowing of Japan"), referring to the relocation of domestic plants in foreign countries such as Korea,

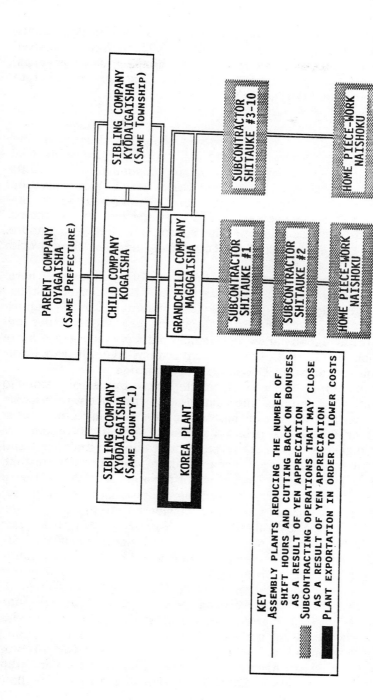

Figure 9.6
Automation and plant exportation for cost reduction in Nakada Sony factories (1988)

Taiwan, or the United States, could have a greater impact now than when the jobs were created in the 1970s. Since the mid-1970s, farming has become clearly subordinated to industry and farmers are dependant upon wage labor income to buy farm machinery and maintain a high standard of living.

SOCIAL EFFECTS

There are three main interrelated social effects of the cutbacks in subcontracting jobs which must be considered together with the cut in rice price subsidies described in Chapter 12. These social effects are (1) increased economic differentiation between households and communities; (2) problems with finding heirs for landless households; and (3) fewer wage labor jobs for females.

Economic Stratification

In prewar Japan large-scale land holdings were a necessary qualification for elected position such as in village government or the Nakada Reclamation Cooperative. Also, one of the few ways for rural landlords to become wealthy was by leasing land to tenants and then buying more land. In the postwar era the relative social and economic importance of land decreased despite the fact that real estate prices increased. The first reason for this was that there was a three hectare upper limit on land holdings until 1970. The second factor was the new income generated by industrial jobs which gradually overshadowed agricultural income. Because the agricultural share of household income decreased, land was often considered in the convoluted system of agricultural economics, to be an over-priced fixed asset of agricultural production. Agricultural land valuation for taxation purposes was figured according to agricultural productive potential rather than its true real estate market value which was driven up because of its commercial potential.

The 1986-87 yen appreciation forced a re-evaluation of the situation. A close look at company employment statistics reveals that there was a strong correlation between land ownership and procuring jobs in the larger companies. The reason for this was based on the preexisting social network of landed farmers who were tied in with industrial development. Similarly, landless households found jobs in the smaller subcontracting companies. This is shown in Tables 9.2 and 9.3.

While there was a wide variation in the amount of land owned by those who were employed by the two largest industries, it is noteworthy that all households with persons employed in the larger companies did have at least some land. Households which did not have any land either before or after the land reform have not fared as well.

The two largest prewar landlords now each own considerable commercial real estate as well as private businesses. In addition to his cab company, the prewar landlord who owned the most land was rumored to have branched out into the sarakin money lending business in a nearby city. The second richest prewar landlord is a leading politician with a son who has a high level job in one of the larger companies.

Should automation force the shutdown of the subcontracting companies, an elite may form on the basis of income obtained by combining farm and factory earnings. The bottom level of landless farmers have not radically changed their position in society due to job insecurity and lower wages inherent in subcontracting work. With three notable exceptions, they still maintain their position on the fringe of hamlet society and will likely be unsuccessful in recruiting heirs to return home to take over the household. They will also experience great difficulty in finding spouses for heirs presently living in the household. If past experience is any indication, these households,

Table 9.2
Land Ownership of K Hamlet Households with Persons Employed in Sony Factories (1983)

Household ID Number	Pre-land Reform Land Owned (ha.)	Present Land Owned (ha.)
13	0	0.4
24	1.40	2.0
26	0	0.6
30	0	0.2
49	6.68	2.3
54	0	0.9
62	0.31	0.8
64	0	0.6

Table 9.3
Landless Non-farming Households and Subcontracting in K Hamlet in Nakada Township (1983)

Household ID Number	Persons Employed	Heir	Type of Job For Employed Family Members
6	1	No	None
9	n/a	No	N/A
12	1	Yes	National civil servant
14	3	No	(1)Electronics (part-time)
			(2)Electronics subcontracting
			(3)Auto/truck repair (self-emloyed)
17	1	No	Land engineer (self-employed)
28	1	No	Sick, moved away, house vacant
35	2	TET*	(1,2)Janitorial work
36	2	TET	(1,2)Janitorial work
38	3	TET	(1)Home piece-work subcontracting
			(2)(3)Electronics subcontracting
42	2	TET	(1)Construction (self-employed)
			(2)Electronics
44	2	Yes	(1)Privately owned coffee shop
			(2)Township office
48	3	TET	(1)Township office
			(2)Electronics subcontracting
			(3)Electronics (part-time)
60	1	No	Hairdresser

*TET (Too Early to Tell); children present but no heir designated.
Source: Author's K Hamlet interviews.

lacking a secure resource base to establish a permanent household, will be more likely to move away from the hamlet.

The Problem of Securing Heirs

In rural Japan it has become increasingly difficult to find heirs to continue the household line and take on the responsibilities of maintenance of the land, household gravesite and ancestral rituals, and responsibilities to non-heirs who have migrated away. This is called the "successor problem" (kōkeisha mondai). The heir problem is two generational. An heir must be selected for the present generation and a spouse must be found for that heir so that offspring can be produced to be an heir for the next generation.

The greatest problem in 1987 was finding wives for male heirs. Nationally there were fewer women than men and according to the 1980 census, there were 8,100,000 unmarried males compared to six million unmarried females between the ages of twenty and forty years. For some rural communities the problem became so severe that they resorted to recruiting brides from abroad. Such was the recent much heralded case where Filipina brides were recruited for the men in Asahi Township of rural Yamagata Prefecture (Satō 1987). The "success" of the Asahi Township case led other townships (Okura in Yamagata Prefecture; Azuka and Yuzawa in Niigata Prefecture; Higashi-Iyayama in Tokushima Prefecture; Masuda in Akita Prefecture, and Sawauchi in Iwate Prefecture) to follow a similar course by importing brides from the Philippines, Korea, Taiwan, China, Indochina, the South Pacific Islands, and Sri Lanka. Even though a bride price of a few thousand dollars was exchanged, there have been several cases of brides unable to tolerate the situation and returning to their homeland (Yamazaki 1987:22). Although the international marriage solution addresses the problem of creating heirs, it has little bearing on the heart of the problem which is the limited and insecure resource base of some households.

The situation for Nakada Township in 1987 is given in Table 9.4, which lists those individuals who are registered in the township consultation program established to procure marriage partners. Individuals or members of their household usually initiate contact with the hamlet representative. Registered young people seeking spouses are also encouraged to attend get-togethers at the agricultural cooperative.

Table 9.4
Individuals Seeking Marriages through the Marriage Consultation
Program in Nakada Township in 1987

	Households Wanting To Receive A Spouse			Households Wanting To Give A Spouse		
Age of Individual	Female	Male	Total	Female	Male	Total
25 and under	55	154	209	117	16	133
26 through 30 years	19	173	192	22	16	38
31 through 35 years	6	127	133	6	15	21
36 and over	1	38	39	0	5	5
Totals	81	492	573	145	52	197

Source: Nakada-chō Kekkon Sodan Shiryō (unpublished 1987 data).

The severity of the problem becomes apparent when considering that in 1987 the township had 3,639 households. Approximately 573 households (15.7 percent) were desperately seeking spouses for their heirs. Almost three-quarters of the persons registered were from households seeking spouses to marry into the household. Of all households seeking spouses, 89.5 percent were seeking brides. This has resulted both from the preference for male heirs and the fact that there was a net out-migration of marriageable young women.

Consequently, the township government introduced a finders fee paid to the hamlet representative for arranging marriages that keep in or attract spouses to the township. The categories that received township funding were: (1) finders fee for any bride (yome) or husband (muko) who would be a spouse for the heir; and (2) finders fee for locating Nakada Township spouses for individuals desiring to marry out of their household and remain in the township after marriage. The finders fee was thirty thousand yen ($240) for individuals who came from outside the township and twenty thousand yen ($160) for individuals who came from within the township.

In some cases, especially for young males with a college education, one way to get a spouse is to find a job in the city and later migrate back to the rural area as the heir. Of course, the main problem is in convincing the new bride to leave the city. Sometimes such brides are reluctant to move to rural areas because of the differences between urban and rural culture. In a 1987 survey conducted by the Miyagi Prefecture Land Commissions, 30.8 percent of the heirs responded that they thought that the rural customs were troublesome. Another 25 percent were undecided as to whether the rural customs were a hindrance (Miyagi-ken Nōgyō Kaigi 1987:19). If the heirs themselves feel uncomfortable, there can be little doubt that spouses feel considerable "culture shock" when moving to the countryside. Such a move also means foregoing one's tenure in the city job and taking a job with a lower entry level wage in a rural company.

As already shown in Table 9.3, the greatest successor problem exists for those households with scarce resources. Households with little or no land and a potential spouse who worked in a subcontracting firm were not as desirable. A bride could fare much better financially in the big city where both she and her potential spouse would be paid higher wages with greater job opportunities. Thus, if the trend continues, it is expected that these households will be unable to find heirs and will have difficulty perpetuating their household lines. A drop in rural industrial labor opportunities as influenced by the yen appreciation would also mean that the service sector might suffer. In such a scenario, even middle-level farmers with land would have difficulties finding brides.

Females May Lose Jobs

During the late 1960s and the 1970s transplanting machines became popular in Nakada Township and females were no longer needed to perform intensive hand transplantation of rice seedlings. At the same time factory jobs appeared and many farm households found a new source of income for their females. Males sought work in the agricultural cooperative, the land improvement district office, the township office, or in the service sector of the economy. Because most of the subcontracting jobs that may be lost are held by a high percentage of young married female workers, a dramatic change in the family farm division of labor and social structure may occur.

The household is primarily concerned with maintaining both agricultural production and industrial employment. This strategy favors the perpetuation of the household over many generations. It seems that individual considerations regarding the household's internal and external sexual division of labor is secondary to perpetuating the household. In 1955 K Hamlet women accounted for 53.9 percent of all farm labor. In 1983, out of seventeen operations necessary to grow rice, women did more work than men in only two areas: the transplanting process (feeding the trays to the men driving the machines) and the hand harvesting (of paddy corners when used in conjunction with men driving the combines). In all other operations men did the majority of the work. In the event that women are laid off from their present subcontracting jobs and do not seek re-employment in other wage labor jobs, the amount of agricultural labor contributed by women will likely increase.

However, most women will first seek re-employment in similar occupations. In K Hamlet only two women had been employed in electronics companies that closed. One of these closings was due to poor labor relations and the other went bankrupt. Both women were successful in finding re-employment in the same occupation. If women begin to experience difficulty in finding jobs in industry, it is expected that the township will make efforts to increase job opportunities in the service sector. However, this may prove difficult because much of the service sector in Nakada Township exists due to the prosperity resulting from the influx of government capital which funds agricultural projects such as the Land Improvement Project and because industrial wage labor has provided a higher standard of living. It is possible that the service sector would also suffer if the subcontracting companies closed.

PRACTICAL STRATEGIES FOR THE 1990s

Government planners in Japan today are confident and have their sights set high on long range plans. In Sendai, the capital of Miyagi Prefecture and regional capital of Tōhoku, the new planning buzzword is Interigento Kosumosu "Intelligent Cosmos," referring to a development plan to plunge the Sendai area into a new era of semi-conductors, superconductivity, and new-age communications. As in other parts of Japan, government planners in Sendai have postponed their plans for the 1980s to the twenty-first century.

It may be more practical for the farmers of Nakada Township to consider practical strategies for the 1990s instead of government goals of the twenty-first century. The part-time farming strategy is at a cross-roads if the yen appreciation, exportation of labor to foreign countries, and deregulation of the rice price support system continue. Most serious, however, is the threat proposed by such industrialists as Kenichi Omae, who desires to take away the family farm inheritance waiver and de-regulate land sales by changing zoning laws. According to Omae this would solve the problem of the urban housing shortage because land prices would drop. This short sighted approach, however, is not taking into account the tremendous social cost that rural Japan would have to bear. This would include a rapid breakdown of the ie system that is the backbone of stability in Japan.

In conclusion, this chapter agrees with the research done by Aoki, Shibuya, and Arai (1987) who have noted that part-time and temporary employees were the first to suffer the effects of the yen appreciation. This chapter also proposes that farming households which own some land have a better chance to remain in the community and find new jobs because they have a fixed base from which to operate. As job security and ownership rights to commercially valued farm land rise in importance in determining class status, increasing economic stratification may occur. Those households that have no resource base may cease to function either financially or because they will not be able to produce an heir to perpetuate the household.

10

The Scale of Rice Farming
and Resistance to the
Land Improvement Project

INTRODUCTION

The problem of scale in Japanese agricultural production has seldom received attention by foreign scholars, although it has been of continuing interest to Japanese (e.g., Hoshi 1975, Isobe 1985, Katō 1981, Kawai 1983(a), Kajii 1982, Yoshida 1975). Foreign pressure to liberalize agriculture has focused Japanese attention on the critical issue of lowering the cost of production to make agricultural products more competitive.

The government directive, "The Basic Direction of Agriculture for the 1980s," assumes that if the scale were larger, Japanese agricultural products would enjoy the benefits of "economy of scale," and become cheaper and more competitive with world prices. At the same time this should reduce the burden on the government, which has subsidized agricultural prices. Towards this end, the Land Improvement Project was introduced in 1966 (see Appendix A). The three goals of the Land Improvement Project are (1) to increase the scale of the paddies from ten to thirty ares; (2) to consolidate fragmented land holdings; and (3) to improve the irrigation and drainage facilities.

The importance of the Land Improvement Project can be assessed by examining the amount of money the government has been willing to allocate. The sum that the Japanese budget devoted to the Land Improvement Project and the Rice Price Support System equals two-thirds of the defense budget in any given year. This fact alone should reveal the significance of an agricultural infrastructure compatible with part-time farming.

199

The Land Improvement Project currently underway is the second land improvement plan. The first Land Improvement Project was carried out between 1900 and 1935 and created rectangular ten are paddies out of the irregular paddies that were based on contours of the natural environment. It also improved drainage and irrigation facilities. The reclamation of Nakada Marsh described in Chapter 3 was undertaken through that reclamation project. As Table 10.1 points out, this project converted about 19 percent of the total paddies in Japan. In Nakada Township, the other paddies were shaped into rectangular form and consolidated in 1951.

This chapter describes how prewar social relations of production were neglected by government policy in formulating the Second Land Improvement Project. It argues that historical rights over land and water were very fundamental to participation in the Land Improvement Project. Last, it maintains that social costs must be considered in making effective government policy.

My thesis is that prewar social relations, which centered around tenant-landlord cultivating rights, were not eradicated but rather transformed by the Land Reform of 1946. The Land Improvement Project, in the context of the rising value of rural farmland, is altering existing land tenure and water allocation rights and furthering the possibility of larger scale farming for the first time since World War II. The Project may well become a battleground for small-scale farmers intent on maintaining their land rights and resisting the situation of becoming full-time wage laborers without land rights. The first part of this chapter will delineate general theoretical considerations concerning the scale of agriculture in Japan. The second part of the chapter recounts the resistance by K Hamlet to the Land Improvement Project.

THE SCALE OF JAPANESE AGRICULTURE

By international standards, Japanese agriculture, with a scale of 1.15 hectares, is one of the smallest in the world. This compares to a European average of 15.3 hectares, or an American average of 175 hectares. Table 10.2 reveals that the cultivation scale of Japanese agriculture has remained constant since the turn of the century. Although there has been recent emphasis on larger farms, the overall trend has not appreciably changed. In view of this, the government plan to bring about a threefold increase in paddy size is monumental.

Table 10.1
Land Improvement Project of 1900-1935

Prefecture	Percentage of Project Completed by Era				Percentage of Paddies Completed
	1900-1905	1906-1915	1916-1925	1926-1935	
Aomori		13.9	32.8	53.7	9.0
Iwate		65.4	13.4	21.8	4.2
Miyagi	4.9	63.0	13.4	19.3	43.7
Akita		14.8	55.5	30.4	4.3
Yamagata		19.3	36.1	44.6	28.0
Fukushima		52.6	22.3	24.9	28.3
Ibaraki	0.2	1.5	49.7	34.6	14.9
Tochigi	0.2	1.9	49.2	31.8	29.1
Gumma	3.4	39.4	23.6	35.2	41.1
Saitama	0.9	22.3	48.3	28.7	36.4
Chiba	0.8	14.3	32.5	52.4	19.5
Tōkyō	0.6	4.4	17.0	73.0	62.5
Kanagawa	3.5	11.3	25.0	59.6	49.6
Niigata	0.4	16.9	31.0	51.5	23.0
Toyama	0.5	13.1	32.5	53.7	8.6
Ishikawa	0.1	12.6	56.0	30.6	39.4
Fukui	0.2	8.0	44.9	45.4	6.7
Yamanashi		37.8	36.0	26.6	10.0
Nagano		14.8	29.7	56.5	12.0
Gifu		11.2	49.8	38.9	12.9
Shizuoka	2.3	8.1	27.1	62.5	28.8
Aichi	1.1	17.4	34.7	47.2	16.3
Mie	0.2	11.9	25.0	51.8	12.0
Shiga	0.7	8.2	27.1	63.8	9.7
Kyōto	1.0	16.1	12.7	70.0	13.3
Ōsaka		1.1	27.6	61.5	12.5
Hyōgo		16.9	27.3	56.6	10.0
Nara		6.7	52.8	40.3	7.9
Wakayama		14.9	50.8	31.0	8.3
Tottori	1.7	23.3	46.5	27.9	25.0
Shimane	0.9	25.0	37.2	37.3	11.6
Okayama		12.3	42.6	46.4	9.3
Hiroshima	0.2	8.9	25.9	66.3	14.8
Yamaguchi	3.2	27.3	29.8	45.2	22.0
Tokushima		31.7	28.2	40.9	11.5
Kagawa		15.3	38.9	46.0	7.7

(continued)

Table 10.1 (Continued)

Ehime		24.3	32.9	43.3	22.2
Kōchi		6.8	15.6	77.2	25.1
Fukuoka		27.5	44.2	28.1	26.3
Saga	0.2	16.1	25.3	59.3	14.1
Nagasaki		37.1	11.2	51.8	10.8
Kumamoto		35.5	26.2	38.1	16.5
Ōita	0.5	12.6	50.9	36.2	10.9
Miyazaki	0.1	50.0	42.0	9.8	25.8
Kagoshima	0.3	28.3	34.0	36.8	36.7
Average	0.9	24.0	33.3	39.7	19.1

Source: Tōbata 1978:228.

In Table 10.2 the area of cultivated land is listed. While cultivation scale has remained fairly constant, land ownership and land rights have undergone change both before and after the war. As described in Chapter 4, during the 1920s peasant resistance to landlords earned peasants the "right" to cultivate land. In fact, this tenant right to land was heritable, and exerted a tremendous influence on rural Japan until the Agricultural Law was revised in 1970, abolishing the tenant right. The 1946 Land Reform likewise affected land rights by turning over tenanted land to the actual cultivators.

LAND RIGHTS AND SCALE

Since 1961 government policy has been more concerned with reduction of production cost and therefore with the scale of agricultural production than with actual ownership or cultivating rights of the land. Government policy ideologically emphasizes the "harmonious development of agriculture and industry" while avoiding social and historical issues which play an integral part of that development.

Government policies often seem inconsistent to the farmers, to whom the rights of ownership and cultivation have meaning based on centuries of social relations. The inconsistencies are not merely due to legal vagaries of the policies themselves, but also to contradictions between de jure and de facto cultivation rights.

Table 10.2
The Cultivation Scale of Japanese Farming Households from 1908 to 1987 (X 1000)

Year	<0.5 ha.	0.5-1.0 ha.	1.0-2.0 ha.	2.0-3.0 ha.	3.0-5.0 ha.	>5.0 ha.	Total
1908	2,016	1,764	1,055	348	163	62	5,408
	37%	33%	20%	6%	3%	1%	100%
1950	2,531	1,973	1,339	208	77	48	6,176
	41%	32%	22%	3%	1%	1%	100%
1960	2,275	1,907	1,406	201	35	2	5,823
	39%	33%	24%	3%	1%	0%	100%
1970	1,999	1,604	1,272	141	55	5	5,176
	39%	31%	25%	3%	1%	0%	99%
1980	1,921	1,304	980	240	82	13	4,542
	42%	29%	22%	5%	2%	0%	100%
1987	1,728	1,181	898	248	93 (1985)	19 (1985)	4,178
	41%	28%	21%	6%	2%	0%	100%

Sources: Hayami (1975:9), Nihon Nōgyō Nenkan Kankōkai (1981:400, 1988:527), Isobe (1982:38)

Note: There are minor discrepancies between these sources. Hokkaidō, with a much larger average cultivating scale, is excluded from these data.

The goal of the Land Improvement Project in Nakada Township was to make agriculture more "rational" (i.e. utilizing scale increases and paddy consolidation). In this way it had much in common with the Land Reform, which gave land to the cultivators. The Land Improvement Project differs from the Land Reform in that the latter emphasized and gave ownership title to the actual tillers, while the Land Improvement Project stressed de jure cultivators, in other words those individuals listed with the Land Commission as being "cultivators." In practice many landlords were registered as "self cultivators" while in reality they were landlords. By doing this the landlords could avoid the restrictions of the Land Commission which legally guaranteed a low rental rate for tenants. Therefore, consolidation of paddies according to the de jure cultivator meant that a de facto landlord with de jure self-cultivator status could have his paddies consolidated for himself rather than for his tenants.

In addition to the problem of land rights, there was a two-dimensional question of cost. First, the project required a large capital outlay on the part of the land owner, and second, once the paddies were enlarged, it would become necessary to buy new equipment because the old equipment was not designed for the larger paddies.

Most farmers welcomed certain parts of the project, such as better irrigation and drainage facilities and the goal of consolidated holdings. Objections surfaced, however, when the policy threatened to take land rights from one farmer and transfer them to another. Thus, it was land rights and not farming per se that was the issue.

REGIONAL AGRICULTURE AND THE INCREASE IN SCALE

Regions of Japan contrast with respect to crop types and scale of agriculture. Government policy has tried to aid areas in which high quality products are being produced at low prices. Regionalization of Japanese agriculture has social consequences not only at the community level but also at the hamlet and household level of production. For instance, areas with more labor-intensive agriculture, such as rice or tobacco, have different social problems than areas which grow land-intensive crops such as wheat or soybeans.

According to the 1980 Agricultural Census, Hokkaidō, with 8.4 hectares per farm, has the largest scale of agriculture in Japan. Tōhoku is second with an average of 1.4 hectares. This contrasts with regions such as Kinki, Chūgoku, and Shikoku where the averages are 0.7, 0.7, and 0.8 hectares, respectively.

Throughout most of the world the scale of agriculture has increased since World War II. In Europe the number of farms over one hectare and under five hectares decreased from 4,030,000 in 1960 to 2,682,000 in 1977, while the number of farms fifty hectares and over increased from 256,000 to 331,000. Farm size in the United States has changed even more dramatically, nearly tripling during the last thirty years. As Table 10.3 indicates, Japan is no exception. In 1955 only 12 percent of the farms cultivated over two hectares while in 1980 this number had risen to over 25 percent. Government policies have forced small-scale producers into non-production and to lease out land but not sell it. Japanese inheritance laws, encouraging farmers to keep the farm in the family, distinguish Japanese small-scale producers from their European and American counterparts. In Europe and America the small-scale farm is usually sold when the household farming business fails. Japanese farmers, who may either retire or move on to another business, usually keep the land in the family and lease it out rather than sell it.

As noted, Hokkaidō and Tōhoku possess a large number of farms over two hectares. In both cases the numbers have steadily increased since 1965. The rate of large-scale farms leasing land is also highest in Tōhoku and has risen from 37.0 percent in 1975 to 54.5 percent in 1980 (Isobe 1982:169).[1]

This diverse scale has special meaning with regard to the amount of labor necessary for agricultural production and time created for off-farm wage labor. This is shown in Table 10.4, which compares the Tōhoku and Kinki Regions during 1960 and 1980, the twenty year time period during which there was widespread dissemination of the rototiller, tractor, transplanter, and of binder mechanization. Kinki, the region located around Kyōtō, Nara, and Ōsaka, makes an interesting contrast because the farmers do not monocrop rice, and have been near industrial centers since the turn of the century. The conclusion that can be drawn is that part-time farming and wage labor in industry go hand-in-hand. As new labor-saving techniques became available, farm household labor saved by machinery became labor that could be reallocated by the household as off-farm wage labor.

When comparing the 1960 and 1980 statistics (b/a), we see the largest decrease in labor in the agricultural labor section,

Table 10.3
The Percentage of Land Cultivated by Farms Over 2 Hectares in Size
(area cultivated by 2 ha. or over/total area in cultivation)

Regions	1955	1960	1965	1970	1975	1980
All Regions	12.0	13.2	15.3	18.8	22.1	25.7
Tohoku	30.8	32.8	33.9	38.2	41.7	46.2
Kanto	13.2	14.4	16.3	18.2	21.0	24.3
Hokuriku	16.2	18.6	20.3	23.2	26.7	30.4
Tosan	3.5	3.7	4.2	6.2	8.0	10.3
Tokai	1.8	3.0	4.0	6.0	7.5	9.5
Kinki	0.5	0.7	1.5	3.4	5.2	6.9
Chugoku	1.9	2.4	3.8	6.5	7.8	9.8
Shikoku	2.4	2.5	4.0	7.9	10.6	12.3
Kyushu	7.6	8.9	11.3	16.7	20.9	24.7
Hokkaido	1.4	0.8	2.2	14.0	32.7	42.7

Source: Isobe et al. 1982.

Note: The figures for Hokkaidō are for the percentage of farms cultivating over 20 hectares.

where the rototiller and rice-transplanter have made an impact. As a result, more time was made available for wage labor in industry (124.5 percent increase). The total labor expended per household was reduced (89.1 percent), but the size of the household had also decreased, causing an increase in per capita labor expenditure. The Tōhoku Region differs from other regions in Japan where rice was less predominant and where there were off-farm job opportunities in the late 1950s and early 1960s. In Tōhoku, industrial development occurred more recently. The timing of this development, which provided off-farm jobs, coincided with mechanization.

There is one important difference between the Kinki Region (the area near Ōsaka, Kobe, Nara, and Kyōto) and the Tōhoku Region. The amount of wage labor in the Kinki Region was nearly the same (98.0 percent) between 1960 and 1980, whereas in Tōhoku, as noted above, there was a 124.5 percent increase.

It is worthwhile noting in Table 10.4 that the average scale differences between Kinki and Tōhoku regions are overshadowed by the similarity of trends among part-time farms of the same scale. Farms under one hectare experienced the most dramatic decreases in the amount of household labor allocated to agricultural production. Small-farms allocated only one-third as much of their labor to agricultural production as they did twenty years earlier because of the introduction of the rice transplanter and other labor-saving machinery described in Chapter 8. Since land is a scarce resource, and not easily obtained, it is difficult to expand the farm scale. This leaves non-farm wage labor as the only viable alternative to increasing household earnings.

The most noteworthy changes in wage labor participation occurred as the scale increased. The fact that the smaller scale farms in 1960 already were engaged in twice as much labor off the farm than on is evidence that the subcontracting system of Japan, over twice as common as in the United States, had solid prewar foundations. At the very least, the high growth rate of the 1960s "Japanese economic miracle" was predicated on the changes in part-time farming and subcontracting that occurred in the 1950s. In addition, the increase in part-time farming should not be associated with farms going out of business so much as farms becoming more heavily capitalized and mechanized enabling them to stay in business, while adding additional income. Large-scale farmers were forced into wage labor to enable them to buy machinery and avoid paying enormous sums for farm labor which was increasingly expensive. Large farms were therefore

Table 10.4
The Scale of Farming and Household Labor Allocation Per Farm in the Tōhoku Region and the Kinki Region (in hours)

Labor Input	Year	<0.5 ha.	0.5-1.0 ha.	1.0-1.5 ha.	1.5-2.0 ha.	>2 ha.
Tohoku Region						
Agricultural Labor (Self-cultivating)	1960(a)	1,497	3,225	4,594	5,390	6,203
	1980(b)	481	1,130	2,311	2,847	3,619
	b/a	32.1	35.0	50.3	52.8	58.3
Private Business (Side business)	1960(a)	536	373	250	269	148
	1980(b)	177	409	163	223	182
	b/a	33.0	109.7	65.2	82.9	123.0
Wage Labor (Industrial)	1960(a)	2,875	1,733	1,194	897	531
	1980(b)	3,579	3,250	3,085	2,540	2,081
	b/a	124.5	187.5	258.4	283.2	391.9
Total Labor	1960(a)	4,967	5,382	6,112	6,628	6,992
	1980(b)	4,425	4,963	5,744	5,841	6,131
	b/a	89.1	92.	94.0	88.1	87.7
Kinki Region						
Agricultural Labor (Self-cultivating)	1960(a)	1,541	3,596	5,980	6,348	7,190
	1980(b)	579	1,471	2,374	2,985	4,711
	b/a	37.6	40.9	39.7	47.0	65.0
Private Business (Side business)	1960(a)	3,211	335	197	160	792
	1980(b)	161	272	382	66	160
	b/a	49.7	81.2	193.9	41.3	173.9
Wage Labor (Industrial)	1960(a)	3,237	1,584	821	666	102
	1980(b)	3,171	3,179	2,749	2,476	1,885
	b/a	98.0	200.7	334.8	371.8	1,848
Total Labor	1960	5,189	5,618	7,126	7,432	7,583
	1980	4,011	5,057	5,668	5,713	6,937
	b/a	77.3	90.0	79.5	76.9	91.5

Source: Kawai 1983(b):32.

able to maintain lower labor input per unit of land than small-scale farms. Table 10.5 shows labor time per ten ares and indicates that the larger farms spent less time producing per unit area and thus benefitted more from the mechanization of agriculture than did small-scale producers.

The smaller farms bought machinery to save household labor because higher income was available from off-farm wage labor. By doing so, they have become more "over-capitalized" than larger scale farms. Farmers also purchased machinery enabling them to farm during the weekends when they had time off from their jobs. Both large-scale and small-scale farming households increased wage labor participation, but for different reasons: the small-scale producers did so to buy machinery to save <u>labor time</u> and farm on weekends, while the larger scale producers did it to save on <u>labor costs</u>.

One modest but notable trend is the declining gap in labor time between small and large-scale farms. This is true in both the Tōhoku and Kinki Regions. However, the disparities between the regions themselves seems to be widening. According to Kawai (1983:27), in 1960 the Tōhoku and Kinki Regions both averaged 175 hours per ten are plot. In 1980 there was a ten hour difference, with Tōhoku having decreased the labor time to sixty-five hours compared with Kinki's average of seventy-five hours. Because of Kinki's lower average scale (0.7 ha. compared to Tōhoku's 1.4 ha.) and proximity to industrial job opportunities, Kinki is more advanced in the "part-time farming phenomenon" than Tōhoku and is a good example of the regionalization process that is occurring in Japan.

COST EFFICIENCY AS VIEWED FROM THE HOUSEHOLD

Making more money off the farm than on it has led to a contradiction. The ideology of the part-time farmer stresses that they are, "farmers who work", rather than "workers who farm." Yet most part-time farmers make more money off the farm than on it. They view the safest and most profitable strategy to be maintenance of agricultural land rights at any cost, even if it means being over-capitalized in equipment and farming on weekends.

Supporting this idea is a notable observation by informants during my interviews. They asserted that their profit per area was higher than large-scale farmers because small-scale farmers can exclude household farm labor costs. This statement is very similar to what Chayanov observed among Russian peasants and supports what Sahlins has argued about the "domestic mode of production" (Sahlins 1972). This is in direct contradiction to government statistics, which indicate that large-scale farms are more efficient because costs per unit of production are lower. Government statistics always treat household labor as if it were wage labor when computing the factors of production. The households themselves make an important distinction between these two. The former is not calculated because the labor relationship (spouse of heir, son or daughter of heir, parent of heir, etc.) is based on single heir inheritance and the land rights guaranteed through it.

THE COSTS OF PRODUCTION

According to government figures, the divergence in the cost of production between large and small-scale farms has grown. This is mainly due to the greater use of machinery suited for large-scale rather than small-scale farming. Supporting the government claim that large-scale farming is more efficient, Table 10.6 reveals that during the last twenty-five years a significant discrepancy has existed between the costs of production for large versus small-scale farms. If the cost of production for farms having a scale of 0.3 hectares or less is indexed at 100, the farms having a scale of three hectares or more would fall at seventy-one.

Before the introduction of power tillers and transplanters, the cost index difference between large-scale and small-scale producers was closer. In Table 10.7, based on a 1958 survey, the cost per unit produced by the farms over three hectares in scale was only ten points lower than smaller scale farms having 0.3 hectares or less.

In Table 10.7 Yoshida (1981:318) has demonstrated that the per paddy costs of production for various size farms were about equal if the household's agricultural wage labor costs were not included in the agricultural production costs. The reader should also see that production costs varied by region. The cost index figures per unit for the Kinki Region in 1958 were higher than for Tōhoku when household labor was included but much lower when household labor was excluded from the calculations.

Table 10.5
Labor Time Spent Per 10 Ares in Rice Production in the Tōhoku and Kinki Regions (in hours)

Tōhoku Region

	0.3 ha.	0.3–0.5 ha.	0.5–1.0 ha.	1.0–1.5 ha.	1.5–2.0 ha.	2.0–3.0 ha.	3.0 or over
1960	225.6	186.4	193.0	183.5	171.2	171.2	140.4
1980	85.2	81.7	74.7	63.6	57.7	57.7	53.2

Kinki Region

	0.3 ha.	0.3–0.5 ha.	0.5–1.0 ha.	1.0–1.5 ha.	1.5–2.0 ha.	2.0–3.0 ha.	3.0 or over
1960	214.0	195.3	175.2	171.2	170.6	N.D.	N.D.
1980	91.6	84.0	77.8	58.2	68.5	59.6	N.D.

Source: Kawai 1983:27.

Table 10.6
Farm Scale, Yield, and Costs of Rice Production in 1981

Farm Scale	Yield Per 10 Ares	Cost Per 60 Kilograms (in yen)	Cost Index
0.3 ha. or less	527	196,171	100
0.3–0.5 ha.	515	179,817	
0.5–1.0 ha.	524	160,020	
1.0–1.5 ha.	538	140,581	
1.5–2.0 ha.	560	128,847	
2.0–3.0 ha.	561	118,003	60

Source: Nihon Nōgyō Nenkan Kankōkai 1988:487.

Table 10.7
Indices of Rice Production Costs in the Tōhoku and Kinki Regions in 1958

Scale of Farm Acreage	Cost Per 10 Are Paddy		Cost Per Unit Produced	
	(-)	(+)	(-)	(+)
Tōhoku Region				
Average	112.0	98.2	106.3	93.2
0.3 or less	95.1	99.2	95.3	99.5
0.3–0.5 ha.	102.3	96.8	98.8	93.4
0.5–1.0 ha.	103.8	98.5	104.6	100.1
1.0–1.5 ha.	109.4	100.0	105.7	96.6
1.5–2.0 ha.	110.4	96.2	106.5	92.7
2.0–3.0 ha.	117.2	98.8	104.6	88.1
3.0 or more	131.1	95.1	115.4	83.9
Kinki Region				
Average	98.2	108.7	100.1	110.9
0.3 or less	118.0	119.6	130.1	131.7
0.3–0.5 ha.	102.3	117.3	105.3	120.7
0.5–1.0 ha.	97.1	109.9	98.8	116.6
1.0–1.5 ha.	96.4	106.2	98.8	108.8
1.5–2.0 ha.	105.6	104.4	102.9	102.2
2.5–3.0 ha.	85.2	95.1	114.5	127.7
3.0 or more	No Data	No Data	No Data	No Data

Source: Yoshida 1981:318.

Note: (-) Household members' (agricultural) wage labor excluded from production costs. (+) Household members' (agricultural) wage labor income included in production costs. The units being compared are for 1 koku, a traditional straw bag of rice.

According to my own research, small-scale producers use a higher percentage of their total production for personal use. There are two types of rice for household use: the first is for household consumption (hanmai) and the second is rice which is given to relatives (zōtomai), city guests and persons to whom a social debt is owed. Combined, the hanmai and zōtomai comprise up to 15 percent of the total rice yield for the average farmer. Hanmai and zōtomai are not taken into consideration when the government constructs charts to show farm scale and production cost ratios and are therefore the data are biased in favor of the larger farms.

The present structure of Japanese agriculture has served as the backbone of the Japanese industrial subcontracting system, which utilizes cheap rural labor. The system of capitalist large-scale agriculture and part-time farming has fostered monocropping, mechanization, higher rent rates, selection of labor saving plant varieties, synthetic fertilizers, and higher rates of capitalization of agriculture. The new plans to enlarge paddy size and foster larger scale farms usher in a new phase of agricultural policy in Japan which, if successful, will force a new relationship between farmers and industry.

THE GOVERNMENT PLAN TO INCREASE FARM SIZE

The government goal is to increase farm size to ten hectares over the next ten years by fostering large-scale farming. This is a continuation of policies that began with the Agricultural Basic Law of 1961 in which land reform laws were gradually lifted. In the government planning publication The Basic Direction of Agricultural Policy for the 1980s (see Appendix A) the Ministry of Agriculture set production goals for 1990. Its plan is to reduce paddy acreage from 2,550,000 hectares in 1980 to 1,970,000 hectares in 1990 and to increase grain feed area from 950,000 hectares to 1,550,000 hectares for the same period. Furthermore, a policy has been established to foster large "core" farms chūkaku nōka that are to possess an average cultivation area of ten hectares by 1990. It is assumed that by doing this the amount of labor time per ten ares will be reduced from the present seventy hours to thirty hours and that the production costs will be halved (Nōsei Shingikai 1982:60).

Although the government has introduced a series of projects to facilitate expansion of production scale, the largest is the

Land Improvement Project, which is attempting to enlarge the scale of paddies from ten ares to thirty ares, improve irrigation and drainage facilities, and consolidate fragmented plots. The Land Improvement Project is attempting to make possible the dual use of paddies so that rice and vegetable crops can be produced on the same field in succession. It also promotes domestic animal husbandry and a supporting fodder and feed system in order to decrease dependence on foreign sources such as the United States, Japan's top supplier.

The surplus of government subsidized rice has become a national issue because of the dual desires for food security and a balanced budget. In 1972 the government (see Appendix A) implemented a Rice Acreage Reduction Policy (gentan) and in 1978 a Crop Conversion Program (tensaku) in order to limit the over-production of rice and promote the conversion of rice paddies into field crops. The Rice Acreage Reduction Policy and Crop Conversion Program are coordinated and facilitated by the Land Improvement Project. These will be discussed in the next chapter.

The funds allocated for the Land Improvement Project and the Rice Price Support System are the two largest items on the Ministry of Agriculture budget agenda. This is key to understanding the LDP policy of fostering part-time farming by paying off the farmers who have disproportionate voting power to pressure for agricultural subsidies and rural industrial employment opportunities. The goal and budget of the Land Improvement Project is given in Appendix A. By 1992 nearly three-quarters of Japan should have completed the Land Improvement Project (Nihon Keizai Shimbun April 1983).

NAKADA TOWNSHIP AND
THE LAND IMPROVEMENT PROJECT

The Land Improvement Project began in 1979 in the southern part of Nakada township, near Hamlet M and downstream on the irrigation line as shown on Figure 10.1. The slope of the land is from upper right to lower left. In such projects it is ordinarily best to start upstream because ideally irrigation projects mirror the topographical slope. When this procedure is followed, problems such as irregular topography and slope of the land can be resolved by down slope fanning out the difficulties of irrigation and drainage line construction. If a

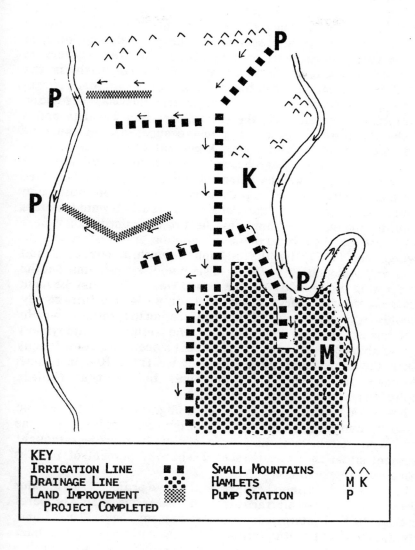

KEY

IRRIGATION LINE	■ ■	SMALL MOUNTAINS ∧ ∧
DRAINAGE LINE	⠿	HAMLETS M K
LAND IMPROVEMENT PROJECT COMPLETED	▒	PUMP STATION P

Figure 10.1
The Land Improvement Project Plan for Nakada Township

project starts downstream, such as in the Nakada case, this limits the potential for an upstream area in formulating its project and, since water flows downstream, presents a problem of what to do with the upstream area drainage.

Nakada Township chose to attempt the project in reverse because government policy stipulated that in order to begin a project, a nearly unanimous consensus of the hamlet residents in the area affected was needed. If under 90 percent of the people were in favor, it might possibly tie the matter up in court for years. According to the 1976 prefectural survey of Nakada Township, 91 percent of the Asamizu residents agreed to undertake the project. According to the supervisor of the local Land Improvement District only 84 percent of the residents in Ishinomori and Uwanuma agreed, splitting up the township on the issue. (Nakada-chō Nōgyōsha Seinen Kaigi 1976:7). When I asked the Hasama District Land Improvement Office about this, they said that they expected the Ishinomori and Uwanuma area residents to change their minds as the project continued.

A more detailed breakdown of opinion was researched the following year (1977) by the Nakada Township Young Peoples Agricultural Association. They distributed a questionnaire to 3,198 farming households in the township. The first obstacle encountered to the survey was that only 44 percent of the households responded, which in itself indicated reserve on the part of many farming households. The results of the survey clearly established that support for the project was considerably less than the prefectural survey indicated. Table 10.8 lists some of the key questions and the responses by the four Nakada Township districts.

The major reason given for opposing the project was that the status quo was satisfactory. Sixty-one percent of the persons who stated that they were against the project thought the present situation was satisfactory while 31 percent of persons had financial reservations.

Large-scale farmers supported the project in greater numbers than small-scale farmers. Of the farms with a scale of less than 0.5 hectares, only 25 percent were in favor. Thirty-four percent of the farms with a scale between 1.0 and 1.5 hectares approved the project as did over half the farms with a scale of over 2.5 hectares.

The project began in the most elevated section of the downstream community of hamlets after the government promised Asamizu a new irrigation pump station and line. With a new

Table 10.8
Nakada Township 1978 Opinion Survey on the Future Intent to Participate in the Land Improvement Project

Question 1: What Do You Want to Do With Your Paddies?

Township Districts	Response 1: Take Part In the Land Improvement Project	Response 2: Improve Part of My Land (Without the LIP)	Response 3: Maintain the Status Quo (Without the LIP)	Response 4: No Comment	Response 5: Other
Uwanuma	26.4	14.7	27.8	20.8	13.6
Ishinomori	15.8	23.4	26.3	18.7	15.8
Asamizu	37.9	21.3	20.0	0	20.8
Takarae	27.0	28.0	22.0	13.0	10.0

Question 2: What Do You Think About The Land Improvement Project?

Township Districts	Pro	Conditional Pro	Con	Not Sure What is the LIP	No Comment
Uwanuma	30.6	9.2	12.4	18.1	30.7
Ishinomori	33.0	8.9	21.8	16.5	19.6
Asamizu	45.2	5.4	11.4	27.8	10.2
Takarae	37.0	7.0	16.0	20.0	20.0

Source: Nakada-chō Nōgyōsha Seinen Kaigi 1978.

pump station the Land Improvement District was able to obtain necessary support for the project because twice as many persons in Asamizu and Takarae listed water shortage as a problem as did the upstream Uwanuma and Ishinomori Districts. The upstream districts also received improved irrigation pump facilities making the upstream water pumping capacity over three times greater than the Asamizu pump facility. Furthermore, the upstream districts fared better when improvements to the irrigation lines are considered. The amount spent for upstream lines was over 2.6 times greater (Kitakamigawa Engan Nakadachiku Tochikairyōku Sōmuka. 1982:7) than that spent for downstream lines. In addition, plans were conceived to construct drainage ditches for the area surrounding the reclaimed marsh. These promised improvements were still not enough to convince the upstream residents to agree to participation in the project. Based on my interviews with 73 households in K Hamlet, the principal reasons for opposition are given below.

UWANUMA DISTRICT AND K HAMLET REASONS GIVEN FOR OPPOSING THE LAND IMPROVEMENT PROJECT

Cost

Costs for the project were calculated to be about 1,100,000 yen per ten are paddy. The national government would shoulder 45 percent of the cost while the prefectural government was to cover 33 percent. The township government would assume approximately 2 percent of the cost, leaving the owner of the paddy to pay 20 percent. This totalled 3,080,000 yen for the average township household with 1.4 hectares and amounted to 69 percent of one year's earnings. In 1982, the year prior to my survey, the average total yearly income for farming households (including non-farm income) in the Tōhoku Region was 4,438,000 yen (Miyagi-ken Nōseibu 1984:170). Understandably a number of the households were disturbed about having to borrow money from the agricultural cooperative at 5.5 percent interest in order to make the payments.

Another cost factor was the loss of 3 per cent of the productive land to construct improved roads and utility facilities. This is significant because farmers viewed the land as a real estate asset. A loss of 3 percent of their land amounted to losing about 10 percent of one year's income for the average farmer.

Disregard for Environmental Soil Differences

The hamlets in the northern part of the township encircle the marsh, which was reclaimed in 1908. Because of the shallowness of the topsoil and the existence of a peat layer close to the surface, farmers remembered transplanting rice by hand in thigh deep mud. Over the years they have tried to rectify the situation by bringing in topsoil from the mountains. This has been so successful that farmers can produce the same grade of rice grown in the more fertile environmental zones. Farmers feared that the project would require large-scale machinery which would mix the peat layer with the thin topsoil layer, thereby lowering the rice grade and yield.

A similar problem was the apportionment of select land and slivers of land created by natural topography and man-made barriers such as roads and utilities. Important negotiations occurred concerning the location of new plots because each plot is different due to varying soil conditions, paddy location, sunlight, water availability, yield, and other factors. The hamlet representatives of the project (kanchiin) facilitated new paddy allocation by giving the Land Improvement Office the rating of each plot on the basis of the above criteria. Nevertheless, some farmers had reservations about whether or not they might receive a paddy of lesser quality in exchange.

Another concern was that during the process of consolidation a household might not receive its fair share of those prized environmental zones having both symbolic and economic value. For instance, in K Hamlet, main households and their branches normally possessed land in the same general vicinity within particular land districts. During the process of household fissioning, plots were subdivided and redivided but each area still holds symbolic meaning for the households involved. The Land Improvement Project, in its efforts to enlarge the paddies, would consolidate these paddies as shown in Figure 10.2.

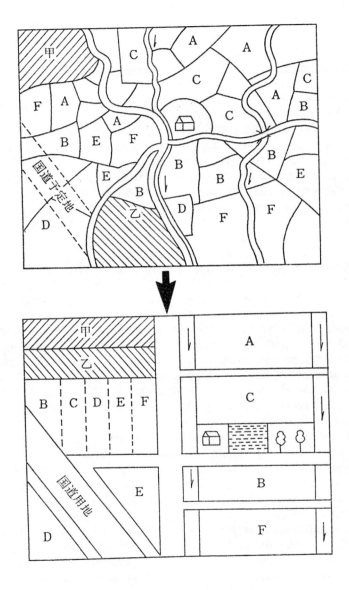

Nōrinsuisanshō Kōzōkaizenkyoku 1982:89.

Figure 10.2
Consolidation of Paddies under the Land Improvement Project

Paddy Recovery Time

As noted above, one objection to the Land Improvement Project was that the peat based soils of the reclaimed marsh or the subsoils in the vegetable areas would mix with the topsoil resulting in lower yields and poor quality rice. Some people also knew of other townships that had experienced difficulties levelling the paddies after completion of the project. Even after plowing flooded paddies, some areas of the paddy might remain higher than others creating irrigation problems, particularly after transplanting. Small transplants at the lower levels of the paddy would weaken by flooding while those at higher levels would not receive adequate water. Generally, it would take five years for a paddy to recover after completion of the Land Improvement Project.

Larger Machinery Would Become Necessary

Due to the formation of larger thirty are paddies, bigger transplanters, which hold more seedlings, would be needed. If the presently used older and smaller machines are used in the newer and larger paddies, the transplanter runs out of transplants at the far end of the paddy where there is no path or road. This necessitates walking to the other end of the paddy where the transplant boxes are stored. One farmer cynically remarked that the real question is whether the planners are creating bigger paddies for the people or for the machines.

The Method of Consolidating Paddies

The Land Improvement Project intended to increase paddy size and attempted to consolidate the registered tiller's paddies. As mentioned previously, the Land Improvement Project, like the Land Reform, is designed to combine the paddies of the actual tiller of the soil. While the size increase met with limited resistance, the biggest debate concerned whether or not the paddy size increase should occur on the basis of existing plot locations (genchika). As described in Chapters 3 and 4, farmers have invested in perfecting their paddy infrastructures and naturally are hesitant about receiving an unfamiliar location. In genchika the paddy size would be increased but the farmers would still maintain non-contiguous plots in different environmental zones. The other alternative, and that favored by the Land Improvement District Office, is to consolidate non-contiguous plot locations (shūdanka).

This debate hinges on the issue of land rights described previously whereby landlords are legally listed as tilling their own paddies. If shūdanka is pursued, the paddies of landlords would be merged to the advantage of the landlords. If genchika is followed, the paddies would be consolidated into a number of non-contiguous groupings.

On the issue of genchika versus shūdanka there is considerable disagreement between the Uwanuma District (upstream) and Asamizu (downstream). In my calculations of the first area completed by the Land Improvement Project in Asamizu, I found the number of non-contiguous paddies being reduced from about seven to three. This is an example of shūdanka, a considerable reduction in groupings. If genchika is pursued in Uwanuma, paddy size will be enhanced but there will be many subdivisions within the enlarged paddies and the number of non-contiguous paddy groupings will be greater. Genchika, therefore, is a standoff between landlords and tenants so that the landlords or large-scale farmers do not get the upper hand (the advantages of scale). One tenant cultivator in Uwanuma, who worked as an agricultural extension agent, informed me of his own dilemma in advocating shūdanka on the job while opposing it in his own backyard.

The Land Commission Favored Large-Scale Farmers

In 1983 the policy of the Land Commission changed to favor large-scale farmers. In the event that a piece of land was to be sold, the first to be notified were the farmers who tilled more than the average for the township, which was 1.4 hectares. Concentric circles were drawn to locate paddies of farmers who met that criterion. A priority list of these larger scale farmers was created to give them the first opportunity to purchase the land. In this way the larger farms were already receiving chances to combine paddies through buying and selling of land.

This may have been one of the reasons that a number of large-scale farmers opposed the project. By stalling and by using existing government policies which give them advantages over other farmers in selecting new paddies to buy or entrust, they would be able to obtain preferable locations for paddy consolidation. If they participated in the project, the new paddy locations would be decided by the <u>kanchiin</u> and the Land Improvement Office. Only 3 out of 13 farms of over 2 hectares in cultivable land have endorsed the project in my research area.

LANDLORDS AND TENANTS

A marked difference of opinion regarding the Land Improvement Project existed between landlord and tenants[1]. Opinion concerning the Land Improvement Project correlated better with prewar landlord or tenant status of the individual than with the <u>dōzoku</u> (<u>edōshi</u>) main or branch household grouping to which the individual belonged. This is true because main and branch households of the same <u>dōzoku</u> were usually in a landlord-tenant relationship. The landlord and tenant data are listed in Table 10.9.

Planners of the project announced that they will wait until a good harvest to push for approval. When I did the survey in 1983 only seventeen households out of the total of seventy-two supported the project. Prewar landlords were four for, two against, and two of no opinion. Of the prewar large-scale tenants, the group that gained the most land in the land reform, three were for and seven against. It is noteworthy that of the three for, each gained more in the land reform than all but one of the seven against. The unregistered lessors of plots did not show any significant preference for or against the project.

It was clear that the major proponents of the project were the prewar landlords and that the majority of farmers who gained land in the land reform were against the project. The leader of the opposition gained 0.6 hectares in the land reform to make a total of 2.9 hectares and is presently the largest cultivator in the hamlet, without leasing unregistered land. In 1988 a new mayor was elected who favored the project and it is expected that a compromise will be sought.

As stated in the section concerning consolidation, it is probable that the large-scale farmers have opposed the project for different reasons than the smaller scale ones. Perhaps large-scale farmers believe that they are already able to lease land and that their cost of production would increase after the project is completed. Small-scale farmers state that they are opposing for

Table 10.9
Prewar Landlords, Large-Scale Farmers, Unregistered Leasing and Attitudes Towards the Land Improvement Project

PREWAR LANDLORDS

Household Number	Land Lost in Reform (ha.)	In Favor or Against the Project
1	1.02	In Favor
2	0.32	Against
4	19.00	In Favor
16	0.43	Against
45	1.60	In Favor
49	3.40	In Favor (major promoter)
53	0.20	In Favor
70	2.72	No opinion

PREWAR LARGE TENANTS (gain of over 0.9 ha. in Reform)

Household Number	Land Gained in Reform (ha.)	In Favor or Against the Project
2	0.93	Against
19	1.90	In Favor
21	1.30	Against
23	0.92	Against
43	1.09	In Favor
46	0.90	In Favor
54	0.90	Against
56	1.10	Against
69	1.23	In Favor
72	0.97	Against

Table 10.9 (Continued)

UNREGISTERED LEASING

Household Number	Prewar Landlord	Amount Leased	Total Owned	A/B Ratio	In Favor or Against the Project
1	X	1.71	1.77	0.97	Pro
3	X	1.45	2.05	0.71	Pro
4	X	1.00	1.10	0.91	Pro
24		0.02	2.00	0.01	Con
37		0.10	0.43	0.23	Pro
45	X	0.90	0.90	1.00	no opinion
49	X	2.10	2.30	0.91	Pro
50		1.70	1.70	1.00	Con
52		0.10	0.55	0.18	Con
53	X	n/a	0.63	--	Con
59		1.10	1.30	0.85	Pro
61		0.30	0.30	1.00	no data
65		0.39	0.55	0.71	Con
70	X	0.16	1.60	0.10	no opinion

financial and ecological reasons. In 1983 when I administered the questions, however, I did not ask about the potential threat to land rights which would undoubtedly be a worthy topic for future investigation. Small-scale farmers possess the machinery to maintain small paddies and make more money in part-time farming than they could if they became full-time workers or full-time farmers. They can only maintain this status if they preserve land rights. As a result, K Hamlet in Uwanuma, an upstream district with prewar local landlord control, contrasts with downstream areas which had a high degree of absentee landlord control.

CONCLUSIONS

Currently, agricultural policy is attempting to change both the scale and ownership pattern of rural agriculture in order to lower the costs of production. While the government does have clear goals about future production costs and prices (see Appendix A), social costs were not considered by this policy. A shift in land rights would mean that many farm households would be forced out of business. Farms have been able to adapt by reducing family size and allocating labor in different amounts to agriculture and industry. For this reason, the number of farming households has remained constant. How long can this continue if the government forces small-scale farmers out of business by giving preferential treatment to large-scale farmers?

The social costs of disrupting rural society where hamlets and households maintain collective identity through communal rights over land cannot be ignored. Japanese government policy needs to give more attention to the social importance of hamlet solidarity based on traditional land use rights. Even within the framework of the existing legislation there is still a possibility of allowing households and hamlets nominal ownership while entrusting land to larger scale farmers in lease agreements.

If small-scale farming households lose their farms and land due to an agricultural policy biased towards large-scale farms, what collective asset or ideology will hold the households together? It is fair to say that the stability of the household system is at stake. Because of the single heir inheritance pattern and household and hamlet members' collective identity with the land, the one thing that has remained fairly constant in

Nakdachō through the "part-time farming phenomena" has been the number of rural households.

Already the process of transforming farmers into off-farm workers has brought with it a decrease of agricultural communal activities. Communal labor exchanges have decreased to a bare minimum. The one remaining hamlet communal labor activity in Nakada, cleaning weeds from the irrigation drainage ditches, will be abandoned after concrete ditches are poured as part of the Land Improvement Project. Land rights constitute the last line of defense for part-time farming households before they are forced to become workers without a land base. Increased resistance to the aspects of the Land Improvement Project that threaten the land rights of small-scale farmers is expected as the Land Improvement Project moves into the more resistant areas of the township.

NOTES

1. Rate of land leasing by large-scale farms is computed by dividing the leased acreage of farms with over two hectares of land in cultivation by the leased acreage of all farms regardless of scale.

2. Resistance to the project is rooted in the historical and social relations of the hamlet and the community. These relations are usually based on kinship ties and household alliances. I have included the case history of one household to show the complexity of the each case and the necessity for examining the rich background behind each paddy.

The Case of Household 57: Branch Household that Opposes the Land Improvement Project

Kayoko Satō's (pseudonym) household was one of several branch households of the Satō main household. The Satō main household was a well-known prewar landlord household. Because Kayoko's household fell into debt in the 1920s, they loaned their daughter to the Chiba main household, a landlord household. At the Chiba residence she did child care and household chores (komori) in exchange for her branch to have the right to till thirty ares of the Chiba's land.

The exchange was fostered by the Satō main household which, prior to this, had arranged one of their daughters to marry into the Chiba household. The Chiba household was reciprocating by taking in Kayoko and giving her branch of the

Satō household land to till.

The rent was only one 60 kilogram bag of rice (per ten ares) on paddies that yielded between five and six bags. This amounted to only a 17 percent to 20 percent rental rate in an era when the rate was sometimes double that number.

Following World War II when land prices and property taxes sky-rocketed, the Chiba main household asked for half of the ownership rights back on these tenanted paddies. This was standard procedure since Kayoko Satō inherited the tenant right from her parents. At the time (1965-1970) many landlords who had such land with registered tenants were anxious to split up the ownership rights so they could demand higher rents (four bags), which amounted to slightly less than one-half the yield.

At the time of the 1983 interview both Kayoko Satō's household (a branch) and the Chiba main household listed their halves as "self-cultivated" in the land office. In the Chiba household case, however, an unregistered tenant did the farming. Both the Chiba and Satō main households supported the Land Improvement Project. Both Kayoko's household (a branch) and the new tenant were against it and were quick to point out that it would cost too much.

11

The Rice Diversion Plan

One of the basic tenets of the Third Comprehensive National Plan of 1978 was "to guarantee a stable food supply." This theme was elaborated in a recent Ministry of Agriculture, Forestry, and Fisheries directive entitled "The Basic Direction of Agricultural Policy for the 1980s" (see Appendix A). This directive promotes the concept of a "Japanese-style diet" (Nihongata shoku seikatsu) in which rice is the principal component. While overall self-sufficiency in grains has dropped to a low of 34 percent in 1985, rice is to be maintained at near 100 percent self-sufficiency. Although rice consumption has declined from eighty-one kilograms per person per year in 1978 to an expected low of sixty-two kilograms in 1990, the amount of rice acreage necessary to maintain 100 percent self-sufficiency is expected to drop from 2,520,000 in 1978 to 1,960,000 hectares in 1990, which is a decline of 22.3 percent (Nōseishingikai 1980:83).

Regulating rice has been a major concern of the Japanese government from feudal to modern times, with the postwar period being no exception. Instituted in 1970, the Rice Crop Diversion Program (see Appendix A) is employed by the Japanese government to control the amount of rice grown, retain farmers' political backing for the Liberal Democratic Party and to subsidize farmers in order to maintain them in the institution of part-time farming. The latter is essential to the overall Japanese development plan.

"Rice politics" and cultural values are central towards understanding the rice crop diversion program. Taken together, the rice crop diversion subsidy and the rice price subsidy constitute over a quarter of the annual budget of the Ministry of

Agriculture, Forestry and Fisheries (1984 Nihon Nōgyō Nenkan Kankōkai:152). The amount of these allotments is a matter of negotiation between farmers and the ruling LDP. The rice price and rice price subsidies increased until 1987 when they started to decline due to domestic and international pressure. Between 1975 and 1983 the total of these two subsidies was maintained between nine hundred million and one billion yen. The amounts are listed in Table 8.10. Together they demonstrate the effectiveness of the farm lobby, the strength of the rural vote, and the determination of the planners to establish a higher rate of food self-sufficiency with rice as the staple food.

The purpose of this chapter is to describe the crop diversion policy and explain how Nakada Township in general and K Hamlet members in particular have dealt with this government policy. Nakada Township tactics devised to address this policy differ from those used by other communities which I have studied (Moore 1985). For instance, in the Shōnai Plain, Toyohara Hamlet took advantage of the government policy by cooperatively trading paddies (use rights only) so that there was a large enough contiguous area of diverted crops to qualify for extra program benefits. Another approach I studied was that taken by the farmers of Hachirōgata Land Reclamation Project, which is the largest land reclamation project in Japan. In this case the farmers have openly defied the crop diversion plan by planting rice, have taken the government to court, and have started to market their rice on the black market. The strategies of Nakada Township present a middle ground and reflect the historical problems of water irrigation and land rights.

K Hamlet farmers initially approached the rice crop diversion program conservatively by individually taking their lowest yielding rice paddies out of production. In its place they usually planted household vegetables. The small paddy size in K Hamlet (ten ares) was effectively utilized since the average acreage required by the crop diversion program equalled the average paddy size (ten ares). After 1987 K Hamlet joined others in the township in a collective action to take advantage of the extra subsidies made available for farmers who traded paddies to create a large contiguous areas of diverted crops.

This chapter demonstrates that social, environmental, historical, and economic factors have contributed to the varied responses to the crop diversion program, despite the fact that the communities share a regional culture and all monocrop rice. The chapter is divided into two sections, the first section concerning the policy itself and the second comprising case studies.

THE RICE CROP DIVERSION POLICY

From the farmer's point of view there have been three stages of rice crop diversion programs. These stages are: (1) the voluntary restriction stage; (2) the mandatory restriction stage; and (3) the mandatory restriction and forced alternative crop diversion stage. The first stage was the voluntary diversion (1969-1971) which represented an alternative subsidy to the high rice price support. By 1969 the government was feeling the effect of its high rice price subsidies, which since the early 1960s had encouraged farmers to grow rice through substantial yearly subsidy increases. Thus the voluntary diversions were a direct result of the high government cost of storing the rice produced through these price incentives. Figures 11.1 and 11.2 show rice consumption decline, increased production, higher rice price subsidies and the elevated quantity of rice being stored by the government. In 1969, under this voluntary diversion program, farmers received twenty thousand yen per 0.1 hectare of rice acreage diverted from rice production. Because this program was not popular, the government was forced to up the incentive to thirty-five thousand yen per 0.1 hectare in 1970. This was desirable to farmers and most welcomed the payment. From the farmers' view this was (and remains) a temporary departure from the pursuit of growing rice. As a result, in 1970, 78 percent of the land diverted in Japan was fallowed while the remainder (22 percent) was converted to growing crops such as vegetables, feed crops, and pulses.

The second stage (1971-1977) consisted of forced restrictions (gentan) where all farmers in Japan had to fallow approximately 10 percent of their rice acreage. This was formally called the Rice Production Control and Diversion Program (Inasaku Tenkan Taisaku) from 1971 to 1975 and the Comprehensive Paddy Field Utilization Program (Suiden Sōgō Riyō Taisaku) from 1976 to 1977. Reduction restrictions were given to each prefecture based on a fixed rate (about 10 percent of the total acreage). The amount received for simple fallowing was reduced to thirty thousand yen per 0.1 hectare fallowed for three years. In order to raise national self-sufficiency in other crops and to encourage a more permanent change, the government offered forty thousand

232

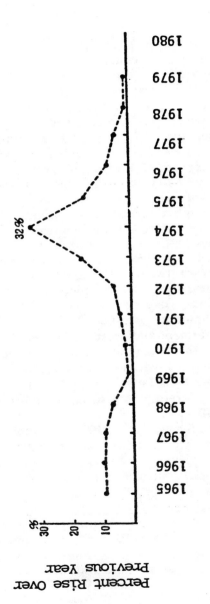

Source: Kome Mondai Kenkyūkai 1981:27, Nihon Nōgyō Nenkan Kankōkai 1965–1981.

Figure 11.1
Annual Increases in the Rice Price

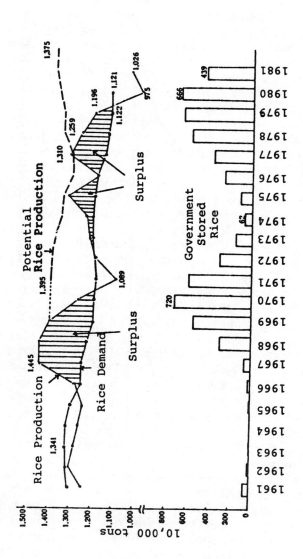

Figure 11.2
The Supply and Demand of Rice in Japan

Source: Kome Mondai Kenkyūkai 1981:27, Nihon Nōgyō Nenkan Kankōkai 1965–1981.

yen per 0.1 hectare to farmers who would divert rice acreage to perennial crops for five years.

For several reasons simple fallowing still proved to be the most popular farming strategy, especially in the rice belt of Japan where the top grades of rice are produced. First, the farmers viewed the gentan program as temporary because it came on the heels of government price incentives for rice (See Chapter 8 for rice prices). Second, there were areas which in the 1950s had just been converted to paddies (called kaiden) and the farmers did not want to spend the time, energy, and money to convert them back again. Third, the financial incentive was not high enough to lure Tōhoku farmers away from their newly developed top rice grades such as the Sasanishiki variety which commanded top market prices.

The third and present stage in the development of the rice crop diversion plan is the use of forced acreage restrictions and mandatory growing of non-rice alternative crops on that land (tensaku). The tensaku policy, as the farmers call the Program to Reorganize the Utilization of the Paddy Fields (Suiden Riyō Saihen Taisaku), builds on the concept of gentan. The 1987-88 tensaku acreage for each prefecture is given in Appendix A.

There are two main dissimilarities between the gentan and tensaku policies. First, the gentan percentage was the same throughout the country. The tensaku policy, however, indulges the rice belt by allowing it a lower percentage of restricted area. Areas such as Hokkaido have the highest tensaku percentage. This has contributed to the regionalization of crop production. Second, in the gentan policy it was possible for farmers to use a simple fallow plan to collect their money. Planting other crops desired by the government was optional. In the tensaku policy, however, farmers were required to shift their restricted acreage to other crops. Simple fallowing was no longer compensated. The subsidy for regular vegetables was 35,000 yen in 1983 plus up to 7,500 yen additional subsidy per 0.1 hectare if the acreage exceeded the minimum required. The appendix includes a more detailed description of the current program and notes that there are incentives to grow specified crops such as soybeans, animal feed, fodder, wheat, buckwheat, and sugar beets as well as long term crops such as orchard crops.

The tensaku stage of rice politics has taken a new twist by giving an incentive to participate in the Land Improvement Project. During the year required to convert one's paddies under the project it is possible to attain tensaku credit and receive subsidies as described below. The government plan to reorganize the paddy fields started in 1966 and is slightly over half completed, with considerable opposition to the project in recent years as described in the last chapter. The government has much at stake in its effort to restructure the vegetable and paddy fields into larger, better irrigated and drained thirty are plots. The combined annual budget totals of the rice price subsidy, the rice crop diversion program, and the project to reorganize the paddy fields amount to 3.4 percent of the annual national budget (as compared to 5.8 percent for national defense).

There are two ways to combine the goals of the tensaku policy with the Land Improvement Project. They are (1) grow a different crop on the land which is in the process of being transformed in the Land Improvement Project, or (2) form a large scale farming group (danchi). The former is a real victory for the farmers since they still have to take the land out of operation for a full growing season so that the bulldozers can restructure the paddies, build new roads, and construct irrigation facilities. The second method, forming a danchi, requires that the effected areas be contiguous with each other, have a scale of between one and three hectares, and grow the same crop (see Appendix A). It is nearly impossible to form a danchi without having completed the Land Improvement Project since irrigation and road width are requirements for forming a danchi. (Irrigation ditches must have a width of over two meters and roads must have a width of at least five meters. This makes it possible for cars to pass each other and for large-scale farm equipment to be employed).

Because of the lucrative danchi subsidies, it is fair to say that the tensaku policy favors larger scale farming.[1] Danchi formation is based on the assumption that a community will participate in the Land Improvement Project. The economic outlay demanded of individual farmers for participation in this project outweighs its benefits from the point of view of some small-scale farmers. Large-scale farming runs counter to the interests of many of these part-time farmers whose income has been greatly augmented through balancing agricultural subsidies with greater rural non-farm employment.

As might be expected, many of these part-time farming households would like to continue maximizing their income. Due to the current style of Japanese rural development, rural income has kept pace with and has even exceeded urban industrial household income. As a result, most farmers wish to remain on the land while working in rural industry and have devised a number of ingenious strategies to maintain their present income and high standard of living.

THE TENSAKU POLICY OF K HAMLET AND NAKADA TOWNSHIP

As a group the strategy of K Hamlet towards the tensaku policy can be characterized as varied methods for preserving the status quo. Nearly all government options are exploited in some manner. The methods are as follows, in order of frequency:

The Kaiden Environmental Zone Approach

The most frequently chosen tensaku acreage consisted of paddy land in the environmental zone which had been converted to paddy use (kaiden) in the late 1950s (see Chapter 2). Because of the late delivery of spring planting irrigation water to this environmental zone, this choice of tensaku acreage had the dual effect of eliminating the last paddy in the planting sequence and therefore shortening the time necessary for planting. As shown on Figure 11.3 almost all the paddies selected were not located in the traditional rice growing areas nor in the former marsh area.

Predictably, farmers chose paddies from less preferred areas, but I was surprised that paddies in the former marsh zone were not selected. I had expected that these paddies would be used because the yields and rice grades there are almost always slightly lower than the other environmental zones. However, a closer look reveals that alternative crops do not grow well or do not grow at all in peat environments such as these. By choosing the higher elevation area which until the late 1950s had been solely used for the cultivation of wheat, oats, barley, soybeans, and mulberry leaves, the farmers selected the area in which the alternative crops would grow best.

This is an indication that other factors such as water allocation and the resulting sequence in spring planting may have been important variables in the decision making process. Both environmental zones that were selected (the high elevation and the river frontage area) represent the last areas to be planted in spring due to water availability. Likewise, vegetable crops were compatible with the sandy soil in these environmental zones.

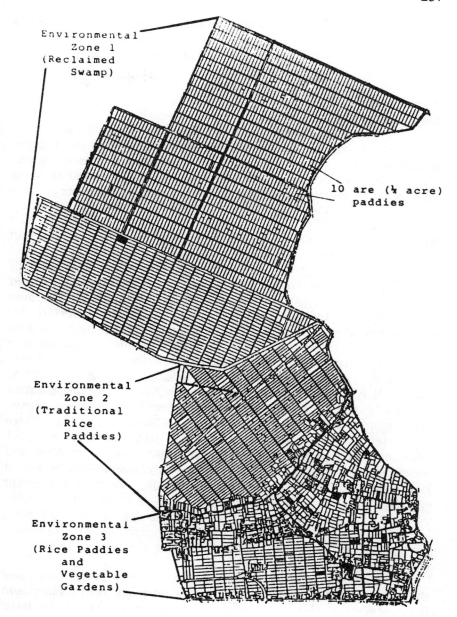

Environmental Zone 1 (Reclaimed Swamp)

10 are (¼ acre) paddies

Environmental Zone 2 (Traditional Rice Paddies)

Environmental Zone 3 (Rice Paddies and Vegetable Gardens)

Figure 11.3
K Hamlet <u>Tensaku</u> Plot Locations

Household vegetables, peanuts, and potatoes were most often grown.

In addition, there were three other advantages to using the kaiden environmental zone. First, the chance of crop damage by early fall frost was reduced. While there are many variables, as a rule the earlier a crop is planted, the earlier it can be harvested. The second, and perhaps more important advantage, was that by eliminating these paddies the farming household could more evenly distribute its household labor over time. Spring planting is labor intensive. This causes a problem because labor intensive periods of farm work conflict with industrial work schedules. The preferred scheduling method was to work outside the farm during the week and work on the farm during weekends. Thus, by eliminating the last paddy planted, it was much easier to reach this goal. An alternative crop was selected on the basis of a growing schedule which did not conflict with the demands of wage labor.

The third advantage of selecting paddies in the higher elevation environmental zone was that houses, buildings, and trees were located there. Rice yields here were lower, the growth rate slower, and probability of frost damage higher due to shading in this location. Paddy names for these tensaku plots reveal proximity to buildings that would create shade: "I no Me" (dialect for "In Front of My House"), "Uchi no Waki"(beside my house), "I No Sho" (dialect for "Behind my House"), "Gongensama Mae" ("In Front of Gongensama Shrine"). Other uses for semi-shaded areas were rice seedling beds (nawashiro), which were replanted to vegetable crops after spring planting. Glutinous rice (mochimai), which is more disease resistant and which sustains a high yield despite shady conditions, was also grown in these areas. However, mochimai did not qualify for tensaku credit.

Paying for the Land Improvement Project

The second method was to sustain all one's paddies and pay the Land Improvement Project thirty thousand yen. The Land Improvement Project in Nakada Township was converting large sections of land in the southern part of the township into thirty are plots, which comprising nearly one-eighth of the township land, were eligible for tensaku benefits. Therefore, the township as a whole had more than its fair share of tensaku land provided that alternate crops were grown on this land.

Thus, several danchi farming groups were formed as private companies under the tensaku law. These danchi farming enterprises also received farm machinery to grow wheat under the programs (see Appendix A). The tensaku program allocated thirty-five thousand yen to persons whose land was undergoing the Land Improvement Project so the thirty thousand yen paid by farmers to have tensaku done for them was paid to the danchi, which also received a bonus subsidy for growing crops desired by the government. These danchi moved from section to section as the Land Improvement Project reached into new areas. As the Land Improvement Project has progressed, this mode of crop diversion has grown in popularity and was quickly becoming the top method for accomplishing the tensaku acreage. Figure 11.4 shows the situation in 1986.

Three reasons were prominent for using paddies involved in the Land Improvement Project for tensaku instead of one's own land. The first relates to the amount of socially necessary rice required by the household. Households with insufficient land often could not afford to diminish their rice yields because the rice was needed for family consumption or for social distribution to relatives. Second, it allowed landlords to continue to receive optimum income from their plots, especially if they were actually leasing plots out to tenants but did not have this fact registered with the Land Commissions. The Land Commissions had an established legal rent rate well below the going black market rate, which was half the production paid in kind. In this case it was more profitable to pay thirty thousand yen to the Land Improvement Project and still receive the four or 4.5 bags of rice rent (each bag bringing over twenty thousand yen). A third reason for using the Land Improvement Project for tensaku was to more fully utilize farm household labor. This was the case in the event that the household had an over-supply of labor power. By using these people to grow rice, household income was being maximized because more time could be spent intensively increasing the yield.

Growing Special Subsidized Crops

A third K Hamlet strategy was to grow government preferred cash crops, which varied with the region of the

240

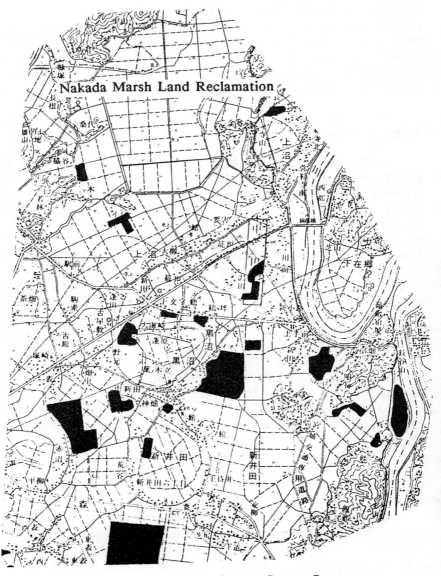

Source: Miyagi-ken Hasama Nōgyō Kairyō Fukyūsho 1987.

Figure 11.4
1986 <u>Tensaku Danchi</u> Locations

country. Some of the crops grown by K Hamlet farmers were soybeans, wheat and animal fodder. These were subsidized by fifty thousand yen which is fifteen thousand yen more than would have been received by simply converting to other vegetables. Another approach was to construct a greenhouse on the land and permanently change production over to a cash crop, such as cucumbers or eels.

Relatives Growing Tensaku Crops for Each Other

A fourth approach was to ask a relative to grow the tensaku crop. In one case, three branch (bunke) households received tensaku credit through their main household (honke), which raised eels. Eel raising, like other greenhouse projects, qualified for the tensaku program. This became a delicate point because of the traditional social obligations between main and branch households and because the environmental pollution caused by the greenhouse heating technique was annoying to neighbors. The main household's greenhouse heating system burned old tires as fuel and these neighbor-relatives were affected by the dense smoke produced. The main and branch households seemed to strike an agreement based on a combination of traditional social obligations, economic considerations, and environmental factors.

Another approach using relatives was followed by two bunke of the same honke, who established an informal leasing system. In this case, one of the bunke households consisted solely of an elderly woman. She leased out her land at low rates to the other branch households who in turn used one of these plots for tensaku.

Cut Early, Collect Subsidies, and Harvest Late

The fifth strategy was a variation of planting rice for animal fodder. In the case of rice grown for fodder, the rice must be cut before a certain government specified date. In this way a farmer could collect the special government subsidy for growing fodder. If there was no early frost, the crop would grow back making it possible to harvest. Since this method was illegal, produced low grade rice, and did not guarantee that all the paddy would grow back, it was rarely practiced.

Village Commons as Tensaku

One last strategy worthy of note was a village level collective method of dealing with the tensaku policy. K Hamlet was part of Sakuraba Village which was amalgamated into Uwanuma Village. Uwanuma Village formerly owned Mount Oizumi as a commons. The mountain was leveled into paddy land and was used to generate gentan and tensaku income. This jointly owned land generated approximately 3.5 million yen of tensaku money which was then divided among the cooperative members. Based on this number, each member household received slightly less than ten thousand yen from the commons. The fifteen Uwanuma hamlets formed a cooperative called Uwanuma Ringyo Kyokai which was run through the agricultural cooperative.

THE CROP DIVERSION PLAN AND
THE ORGANIZATION OF FAMILY FARMING

The tensaku policy is a government attempt to decrease the costly burden of storing excess rice by lowering the production of rice to equal current consumption levels. Additionally, it directly attempts to boost participation in the Land Improvement Project and to expand farm size to form five hectare core (chukaku) farms.

Enlarging farm size does not necessarily mean that small-scale farm owners have to sell out because there is a viable alternative. One alternative suggested by the government has been that ownership rights should be sustained while use rights are transferred. In this way hamlet unity and hamlet usufruct rights can be retained.

As shown in Chapter 4 an obstacle to this is that memories of strong prewar tenant rights accompanied by the idea of "compensation" to tenants who lose these rights to the land is still very real. If small-scale producers decided to abandon farming, keep their land ownership rights while making lease agreements with large-scale farmers, they would still worry about what might happen if Japan's "industrial miracle" collapsed. They would worry that if the large-scale farms gained use rights to the land, the small-scale owners might have a difficult time reclaiming the right to farm it.

Much of rural Tōhoku hamlet life centers around cooperative work and individual and hamlet usufruct rights over land. This is because the rural farm household and the hamlet of which it is part counts land and household labor as its principal assets. As the average farm size expands due to policies such as the crop diversion program and the Land Improvement Project, it is probable that smaller farms will sell out or be forced out. It is clear that these government policies favor the larger scale farms in order to lower production costs but without consideration of the social impact on use rights vis-a-vis private ownership rights. Hamlet life is based on social relations, and shared rights, obligations and responsibilities of small-scale farming households. These are reinforced through group values and usufruct rights on hamlet land.

Accordingly, when one speaks of "the decline of family farming" or whether or not Japanese agriculture can survive (Ogura 1982), one must consider the values and social relations, shared rights over land and labor, and obligations and responsibilities that are inherent to its survival. Certainly the old traditions are changing. In K hamlet and in Nakada Township old values and social relations have resurfaced in new ways. In K Hamlet the local landlords, who also were typically from main households, lost much of their power base (land) in the land reform. However, the land reform was only directed at the quantity of land owned so the landlords held on to the land nearest the hamlet. This land has taken on a new meaning in the context of recent rural development because it happens to be most suitable both for _tensaku_ and for industrial development.

K Hamlet took the "safe" way out and used diversified _tensaku_ methods. Small-scale, fragmented paddies in various environmental zones, in fact, aided the hamlet in its approach to _tensaku_. Almost all households could identify a "least desirable" paddy to take out of rice production and on which to grow a vegetable crop.

In a survey concerning K Hamlet's future farming orientation, 72.5 percent of the hamlet stated that they desired to preserve the status quo, 10.3 percent wanted to reduce their holdings, and 12.1 percent wanted to increase the scale of farming (Nakada-chō Sangyōka 1983). Small-scale farmers will continue to oppose giving up their land and may thwart the state's effort to create a "rational" agriculture. In this sense, I do not see a significant "decline of the family farm" in K Hamlet. A course towards scale expansion and full-time farming, however, would eventually lead to the destruction of the land right base of the household and for every farm that increases its land base to five hectares there would be four that lost it. Industrial work combined with part-time farming allows the household to keep its land rights and perpetuate itself. Government programs which promote the loss of land rights might spell the beginning of the end to hamlet and inter-village cooperation as a gap forms between farmers and non-farmers and between "big" farmers and "small" farmers.

Will heirs be found? If land rights are lost what would be the advantage of remaining in rural areas where the industrial pay is lower than that in the city? If farmers entrust their land to other hamlet members and retain title rights, it may be possible for hamlets to retain their cooperative spirit. At present it appears that part-time farmers will not give up their land although some indicate they should entrust it (keep their ownership rights while giving another farmer use rights for a stipulated time period).

Subsidized rice prices and crop diversion subsidies are essential in sustaining the quality of hamlet life that is based on the land rights of its members. If either the subsidized rice prices or crop diversion subsidies are suddenly dropped or if there is a sudden widening of the framework for free trade of foreign agricultural products, it is easy to conceive of a situation similar to Taiwan which has been characterized as "agricultural degradation" (Huang 1981). In Huang's Taiwanese case, there had been a decline in the number of persons involved in agriculture, a decrease in the rural population's income compared to the urban population's income, and an inability of the rural population to cope with new circumstances.

Taiwan and Japan contrast both in the type of household social organization and the type of national agricultural policy and its relation to industry. In the Japanese case, the number of farming households has not declined to the extent that it has in Taiwan although the number of persons farming has decreased. Unlike Taiwan, the income that Japanese farmers now enjoy is due to successful rural industrial development alongside subsidized agriculture. Industrial development has proven successful in part because it has done a remarkable job at balancing rural and urban income. Rice price and diversion subsidies have played an important role in addition to rural industrial employment opportunities towards achieving rural-urban equality seldom seen in other industrial societies.

NOTES

1. By "large-scale" I mean that it is large compared to the Japanese experience and not by international standards.

12

The Future:
Domestic Agricultural Forecasts
and the GATT Negotiations

INTRODUCTION

The future of Japanese farming rests on the outcome of the free trade question being debated by domestic and international policy makers. Attacks on the farm sector of the economy have come from both inside and outside Japan. Criticism has been mainly directed at rice production, which is protected from competition by foreign imports while being heavily subsidized through government price supports.

Rice, Japan's main crop, is grown by about 80 percent of Japanese farmers. It constitutes about one-third of the agricultural product and 28 percent of Japanese caloric intake. Domestically, industrialists led by the Japan Federation of Economic Organizations (Keidanren)[1], have lobbied to lower agricultural production costs through larger scale farms, take control of the rice distribution system from Nōkyō, lower the amount of rice price support, and import rice for industrial processing and special categories of table rice (Keizai Kōhō Center 1987). According to a survey conducted by the Housewives Federation in 1986, over half of the housewives in Tokyo were in favor of imports under certain conditions (Furuzawa 1988:91). The United States Rice Millers Association (RMA) and the Rice Council for Market Development have claimed that Japan's rice policy violates international trade rules by denying fair and equitable access to the Japanese market. The United States side has also explored the issue from the standpoint of the Fair Trade and Competitiveness Act of 1988.

This chapter argues that the debate over free trade of rice imports to Japan is an international rather than a bilateral issue and that free trade on agricultural commodities should be resolved by the General Agreement on Trade and Tariffs (GATT) and the Organization for Economic Cooperation and Development (OECD). Moreover, in view of the free trade agreement between the European Community (EC) countries which will go into effect in 1992 and the 1988 Canada-United States free trade agreement, it is clear that Japan will also have to play an expanded international role in shaping the future of international agricultural trade. In this chapter the RMA complaint is considered with regard to American agricultural export policy. A summary of Japanese agricultural planning goals for the future is also reviewed. This is followed by an examination of the effects of changes in Japan's farm policy in general and specifically its rice policy.

UNITED STATES RICE MILLERS ASSOCIATION PETITION

The RMA and the Rice Council for Market Development have claimed in a petition filed on September 14, 1988 under Section 301 of the 1974 Trade Act, that Japan's rice policy violates international trade rules by denying fair and equitable access to the Japanese market. This marked the second time the RMA filed such a complaint immediately prior to an important election. Encouraged by the impact of the first complaint filed just before the 1986 Senate elections in California, the second complaint was endorsed by President Bush in his bid to win California and the South in the 1988 presidential election.

The RMA suit filed with the United States Trade Representative (USTR) asks Japan to open its rice market, with a phase in quota of 2.5 percent of its annual rice consumption each year for four years. By the fourth year this would amount to ten percent of the Japanese market. The RMA believes that it has an unassailable legal case which can be presented to GATT. In November 1987 GATT ruled that Japanese import restrictions on ten groups of processed agricultural commodities and food products violated Article 11, which prohibited the use of quotas and other measures to restrict imports.

According to the RMA, by the end of the fourth year 10 percent (approximately 1 million tons) of the rice consumed by Japanese would be imported. In 1988 Japan was importing about

0.2% of rice consumed. At current rates of production it would be difficult for California to produce much more than 10 percent of Japan's consumption. However, the production of medium and short grain rice from all states could potentially supply twice that amount but it is unclear if the Gulf States long-grain rice production could be shifted to medium and short-grain rice production if the market opened. Tables 12.1 and 12.2 give United States rice production statistics.

Long grain rice grown in the southern states comprises the largest segment of United States rice production, as illustrated by Tables 12.1 and 12.2. California accounts for about 54 percent of the total production of short and medium grain rice, the type which is exported to Japan. How much short and medium grade rice could be produced by the United States depends on how much long grain rice acreage could be shifted into medium and short grain production and if this rice would be of a marketable quality.

The assumption of the RMA is that United States medium grade rice such as the Botan, Kokuhō, and Nishiki varieties in California could be sold in Japan as standard table rice grade (hyojunmai) more cheaply than Japanese rice. In fact, the retail price of U.S. No. 1 Fancy Botan rice in a Little Tokyo grocery in Los Angeles was $7.35 for twenty-five pounds on January 23, 1989 and $14.80 for fifty pounds in Columbus, Ohio on February 5, 1989. As of April 1, 1989 the government regulated consumer

Table 12.1
1988 United States Rice Acreage, Yield, and Production by State

State	Area Harvested (1,000 acres)	Yield (pounds)	Production (1,000 cwt)
Arkansas	1,170	5,250	61,425
California	420	7,100	29,820
Louisiana	521	4,550	23,706
Mississippi	270	5,300	14,310
Missouri	77	5,100	3,927
Texas	398	5,900	23,482
United States Total	2,856	5,486	156,670

Source: USDA Economic Research Service 1988:24.

Table 12.2
1988 United States Rice Acreage Planted by State

Grain Type	State	Area Planted (1,000 acres)
Long Grain		
	Arkansas	1,044
	California	50
	Louisiana	365
	Mississippi	275
	Missouri	74
	Texas	388
	United States Subtotal	2,196
Medium Grain		
	Arkansas	135
	California	330
	Louisiana	160
	Missouri	4
	Texas	12
	United States Subtotal	641
Short Grain		
	Arkansas	1
	California	45
	United States Subtotal	46
United States Total		
	Arkansas	1,180
	California	425
	Louisiana	525
	Mississippi	275
	Missouri	78
	Texas	400
	United States Total	2,883

Source: USDA Economic Research Service 1988:24.

price for standard level first grade rice in Japan was 18,487 yen for sixty kilograms which included the 1.5 percent decrease in the rice price and 3 percent increase due to the new consumers tax. Thus the price of $0.30 per pound in Columbus was about 3.7 times cheaper than the $1.12 per pound price in Japan.

However, as Kazushige Kawai (1987:32) has pointed out, shipping the rice to Japan may make the final price somewhat higher than expected. Kawai notes that the transport cost (unchin) and insurance fees (hokenryō) in getting the rice from an American port to a Japanese port is roughly 10 percent of the American export price. Conveying the rice to regional markets incurs tariffs (kanzei) which currently average 9 percent for imported agricultural goods. To this must be added the enroute administrative fees (ryūtsūkeihi) of about 20 percent which include transportation, storage, personnel expenses, and interest. On top of this must be added the handling fees (tesūryō) charged by the routing companies (ryūtsūgyōsha), which include trading companies, wholesalers and retailers. In 1988 this amounted to over 20 percent for independently routed rice. Thus, if we started with a $0.30 per pound retail price and reduced it by 20 percent to get a wholesale price, the resulting figure would be about $0.24. After adding the extra costs noted by Professor Kawai, the Japanese consumer price would be $0.38 making the real price differential with the United States only 2.9 times cheaper instead of 3.7 times cheaper as noted above.

Cheap Is Not Cheap, Expensive Is Not Expensive

A Chinese proverb "Guide bugui, jiende bujien" (What appears to be expensive is not expensive, what appears to be cheap is not cheap[2]) sums up the situation of cost comparisons in the international rice market. Table 12.3 gives a production cost comparison between Thailand, the United States, and Japan. As is clear from the table, Thailand's cost per ton is considerably lower than that of the United States. Likewise, the average 1988 Rotterdam quotations for Thai rice were $359 and $318 per metric ton for SWR 100% Grade A, bulk and SWR 100% Grade B, bulk compared with $408 for United States No.2 milled rice. Looking at this, one might wonder why Thai long grain rice did not undersell American rice on the world market. It is clear from Table 12.4 that Thailand exported about twice as much rice as the United States. Table 12.4 shows the world production of rice and Table 12.5 lists exports and imports. The United States has only been able to keep its share of the world market through export targeting. About 46 percent of all rice exported from the

Table 12.3

1984–1985 Cost per Hectare for Rice Production in Thailand, the United States, and Japan (in yen[a])

	Thailand	California	Texas and the South Gulf	Japan
Seed and Transplanting	928	1,830	2,477	5,629
Fertilizer	36	3,201	4,338	22,149
Pesticides	1,109	1,758	4,059	14,496
Heating and Lighting	67	5,535	9,819	8,370
Other Materials	0	NA	n/a	4,654
Irrigation[b]	428	1,559	2,595	11,442
Rent	1,424	7,601	8,271	17,826
Buildings and Land Improvement	0	n/a	n/a	8,701
Machinery and Equipment	493	5,016	9,078	84,750
Draft Animals	3,688	0	0	0
Labor (Household and Hired)	10,723	2,393	4,894	110,191
Rice Polishing	n/a	n/a	n/a	n/a
Secondary Products	n/a	0	0	12,932
Sub-Total	18,896	28,893	45,531	275,276
Interest (Fixed and Variable Capital)	3,281	1,528	2,939	15,379
Land Rent	4,846	9,115	3,686	63,135
Total Cost	27,023	39,536	52,156	353,790
Yield (tons white rice/hectare)	1.18	5.50	4.25	4.92

a The conversion rate used was 1 baht= 6 yen and $1.00=155 yen.
b Irrigation costs are only those incurred as user's fees. Actual costs may be substantially more.
Source: Kano 1987:105.

Table 12.4
<u>World Rice Production in 1987-1988 (in million metric tons)</u>

<u>Country</u>	<u>1987-1988 Production</u>
Bangladesh	23.0
Burma	12.2
China, Mainland	174.4
India	79.5
Indonesia	38.7
Japan	13.3
Korea, Republic of	7.6
Pakistan	4.8
Thailand	17.3
Subtotal	370.3
Australia	0.8
Brazil	11.0
EC-12	1.9
All Others	63.3
Total Non-United States	447.3
United States	5.8
World Total	453.1

Source: USDA Economic Research Service 1988:33.

United States received some kind of export subsidy to either lower its price or make purchase of it easier through low interest loans (USDA Economic Research Service 1988:36). In 1985 the Food Security Act authorized $1.5 billion to fund an Export Enhancement Program (EEP) through 1988 to make up the difference between the domestic market price and the international price. This act also reimburses export promotion expenses through the Targeted Export Assistance Program (TEA) and has been expanded to a maximum of $2.5 billion through 1990. Because of this act, Thai rice producers were forced to keep their prices low to compete with subsidized American rice.

Table 12.5
World Exports and Imports of Rice in 1988 (in 1,000 metric tons)

Exporting Countries	Amount	Importing Countries	Amount
Argentina	160	Bangladesh	350
Australia	500	Brazil	25
Burma	100	Canada	125
China, Mainland	700	China, Mainland	300
China, Taiwan	125	Cuba	200
EC-12	1,065	East Europe	280
Egypt	100	EC-12	1,092
Guyana	35	India	700
India	200	Iran	450
Indonesia	0	Iraq	650
Japan	0	Ivory Coast	200
Korea, D.P.R.	250	Korea, Republic of	0
Nepal	0	Kuwait	90
Pakistan	1,000	Malagasy	70
Thailand	4,100	Malaysia	350
Uruguay	200	Mexico	0
United States	2,200	Nigeria	300
Vietnam, Soc. Rep.	50	Peru	200
Other	380	Philippines	180
		Saudi Arabia	500
World Total	11,165	Senegal	350
		South Africa	175
		Sri Lanka	200
		Syria	125
		U.A. Emirates	150
		USSR	150
		Vietnam, Soc. Rep.	400
		Other	3,463
		Unaccounted	90
		World Total	11,165

Source: USDA Economic Research Service 1988:33.

How "Free" Is "Free Trade"?

The RMA complaint against Japan is grounded in GATT Article 11 "General Elimination of Quantitative Restrictions." Paragraph 1 of Article 11 states:

> No prohibitions or restrictions other than duties, taxes or other charges, whether made effective through quotas, import or export licenses or other measures, shall be instituted or maintained by any contracting party on the importation of any product of the territory of any other contracting party or on the exportation or sale for export of any product destined for the territory of any other contracting party.

(General Agreement on Tariffs and Trade 1947)

Japan, having joined GATT in 1955, is committed to agricultural trade liberalization in the long run. The problem is when and how. Thus far Japan has been exempted from free trade on rice because the government regulates rice and because Japanese food security is tied to rice.

Perhaps equally important is the issue of farm subsidies. GATT Article 16 specifically cautions against the use of subsidies, "including any form of income or price support, which operates directly or indirectly to increase exports of any product from, or to reduce imports of any product into, its territory..."

Both the United States and Japan heavily subsidize agriculture in different ways. Japan spent about $9 billion in farm subsidies in 1986. The United States spent $26 billion in 1987 (Jones 1987:11). Rice subsidies are estimated to be around $800-900 million depending on the method of accounting used. The EC form of agricultural subsidies, the Common Agricultural Program (CAP), utilized $23 billion in 1986 causing economic hardship for the EC. The use of tariffs has been opposed by the Cairns Group countries who assert that they do not use any subsidies at all. These countries include Argentina, Australia, Brazil, Canada, Chile, Colombia, Fiji, Hungary, Indonesia, Malaysia, the Philippines, New Zealand, and Thailand.

In Japan the rice price and the diversion program described in Chapter 11 are the main beneficiaries of subsidies. Japan also subsidizes agricultural commodity prices for wheat, barley, soybeans, and milk for processing. Stabilization prices are maintained for pork, wagyu beef, and dairy beef. Three major items were subsidized in 1986: income and price stabilization at

526,813 million yen; rice price supports at 363,800 million yen; and the rice diversion program at 232,400 million yen. These totaled about $9 billion.

United States farmers receive subsidies under the Food Security Act of 1985, the mainstay of the United States agricultural export strategy. This strategy combines subsidy programs and intensified marketing efforts by commodity producer associations. The three main export credit sales programs used in the United States during 1989 were the GSM-102, GSM-103, and PL 480. GSM-103 was established under the Food Security Act to guarantee repayment when private commercial credit was given to selected third world countries. GSM-102 did the same for short-term credit. Public Law 480, initiated in 1954, has provided credit at low interest rates with terms of up to forty years. The recipient government was responsible for selling the commodity on its domestic market and the debt could be forgiven if the recipient government achieved certain agreed upon development activities. Table 12.6 shows United States rice production and the large amount of rice that is exported.

Public Law 480 in the postwar recovery period is an example of farm subsidies. Bread baked with subsidized wheat was used in the Japanese school lunch program. Advocates note that it was during this period that a ten year dairy promotion program was started, along with increased importation of livestock feed contributing to greater use of meat and dairy products in the Japanese diet. Critics, however, blame PL 480 for the gradual control over the Japanese diet taken by the United States. They believe that the school lunch program was in part responsible for the decline in rice consumption because school children acquired a taste for foreign food. Rice consumption declined from 115 kilograms per capita in 1960 to a mere seventy-five kilograms in 1985. Likewise, American dumping of surplus grain put Japanese domestic wheat farmers out of business in 1959 or forced them to raise other crops such as rice, contributing to the rice glut of the 1970s.

Japan, on the other hand, viewed its rice surplus as an internal problem. It did not sell its rice surplus on the international market during the 1970s. Rather, storage of the surplus was financed through the Food Control Act. By 1985 government funding for storing the surplus decreased and the rice production subsidy fell to 75 percent of that given in 1975 (Nihon Nōgyō Nenkan Kankōkai 1987:164).

In order to get a clear picture of "free trade," a complex system of agricultural policies has to be considered. These range

Table 12.6
Trends in United States Rice Production

Year	Harvested Yield (1000ha)[1]	Yield[1] (kg/10a)	Total Output[1] (1000t)	Exports[2] (1000t)	Demand[2] (1000t)	Domestic Yearend Stocks[2]	Farm Price ($/t)[1]
1965-66	725	477	3,460	1,418	1,081	217	114
1970-71	734	518	3,801	1,461	1,308	611	118
1971-72	736	528	3,890	1,804	1,309	372	148
1972-73	736	526	3,875	1,726	1,324	167	304
1973-74	878	479	4,208	1,604	1,349	255	247
1974-75	1,024	497	5,098	2,194	1,496	232	184
1975-76	1,140	523	5,824	1,732	1,394	1,205	155
1976-77	1,004	523	5,244	2,097	1,618	1,274	209
1977-78	910	494	4,500	2,270	1,248	879	180
1978-79	1,202	501	6,039	2,431	1,708	1,014	231
1979-80	1,161	516	5,986	2,706	1,794	841	282
1980-81	1,340	495	6,629	3,028	2,113	545	200
1981-82	1,535	540	8,289	2,683	2,247	1,602	179
1982-83	1,320	528	6,969	2,219	2,049	2,303	193
1983-84	878	515	4,523	2,272	1,794	1,481	177
1984-85	1,134	555	6,296	1,960	1,911	2,043	144
1985-86	1,008	606	6,122	1,885	2,079	2,483	
1986-87	963	632	6,097	2,576*	2,152*	2,155*	76-94*

1 Unhulled rice
2 Polished rice
* Projected

Source: Kanō 1987:112.

from national agricultural services and infrastructure measures, which usually do not distort international trade, to stabilization programs, indirect income support programs, and direct trade measures such as tariffs, which frequently have direct impact on international free trade (Miner 1988). Table 12.7 illustrates the range of policy measures which directly and indirectly affect agricultural free trade.

Many countries, such as the United States and Japan, utilize a combination of the above. For example, the United States Payment-In-Kind (PIK) Program and the Japanese 1978 Program to Reorganize the Utilization of the Paddy Fields (see Appendix A) both incorporated acreage output reduction programs (stabilization programs) with indirect income supports. The real question to be decided at the GATT Uruguay Round is where to draw the line between measures which directly and indirectly affect free trade. Recently the debate has centered on the concept of "decoupling" (Miner 1987). Decoupling refers to the separation of farm programs from production, consumption, and trade in the manner shown in Table 12.7 so that a distinction can be made between those policies which have great or little impact on production and markets. By international agreement countries could decide which type of policies to use.

The argument presented in this chapter is that long-term cultural considerations must take priority over short-term stopgap measures. For example, environmental conservation and protection of rural social life have long-term consequences for every economy. Efforts must be made so that each society can protect its cultural adaptation to the natural environment as well as cultural integrity with respect to ideology and social organization. This is a long-term cultural adaptation factor. The importance of long-term cultural factors such as food security and environmental protection were noted in the OECD Communique on Agriculture in Paris on May 13th, 1987 although the overall message of the communique was to promote free trade.

AGRICULTURAL TRADE RESTRICTIONS IN JAPAN, THE UNITED STATES, AND THE EC

One of the reasons why Japan was singled out for criticism was its large number of restricted items. Table 12.8 lists the twenty-two major items restricted in 1987. In addition, some

Table 12.7
Agricultural Policy Instruments and Free Trade

Indirect Affect on Free Trade   Direct Affect on Free Trade

General Examples	National Agricultural Services	Framework Measures	Stabilization Measures	Indirect Income Support Programs	Direct Trade Measures
	research, development and extension	rural development	income stabilization	price and market support, deficiency payments	border measures
	technical and health standards	infrastructure development	stockpiling and buffer schemes	administered pricing, non-commercial state trading	export assistance
	grading and inspection standards	conservation programs	market flow controls	production subsidies	export subsidies
	farm tax and financial services	farm loans	resource inventory	input subsidies (commodity specific)	
	emergency/disaster programs	national direct income transfers	marketing advances and guarantees		
		international development assistance	commercial state trading		

Source: Miner and Hathaway 1988:104-105.

Table 12.8
Japan's Residual Import Quotas in 1987

CCCN Number	Description and Examples
1. 02.01-1 ex	Beef and veal, fresh, chilled or frozen, and offals, except tongue and internal organs
2. 03.01-2-(2) ex	Herring, cod (including Alaska pollack) and its roes, yellowtail, mackerel, sardines, horse mackerel and saurel, fresh, chilled or frozen
3. 03.02-1 ex	Hard roes of cod (including Alaska pollack), dried, salted or smoked
03.02-2-(1) ex	Cod (including Alaska pollack), herring, yellowtail, mackerel, sardines, horse mackerel, saurel and niboshi, dried or salted
4. 03.03-2-(1) ex	Scallops, adductors of shellfish, cuttlefish and squid (other than Mongo ika), fresh, chilled or frozen
03.03-2-(2) ex	Scallops, adductors of shellfish, cuttlefish and squid (other than Mongo ika), dried or salted
5. 04.01 ex, PL	Sterilized or frozen milk and cream and other cream with a fat content of 13 percent or more, fresh, not concentrated or sweetened
*6. 04.02	Milk and cream, preserved, concentrated or sweetened
*7. 04.04-1 L	Processed cheese
04.04-2 ex	Other cheese (except natural cheese) and curd
*8. 07.05-1	Small red beans, dried and shelled
07.05-2 ex	Broad beans, horse beans and peas, dried and shelled, except seed

Table 12.8 (Continued)

07.05-4 ex	Other beans, dried and shelled, except green beans and seed
9. 08.02-2 ex	Oranges, fresh
08.02-4 ex	Tangerines, fresh
10. 08.11-2 ex	Oranges, provisionally preserved by sulfur dioxide gas or other preservative gases
08.11-3 ex	Tangerines, provisionally preserved by sulfur dioxide gas or other preservative gases
11. 11.01-1	Wheat flour
11.01-2 ex	Rice flour and barley flour
12. 11.02-1 ex	Groats, meal and similar preparations (except germ) of wheat and rice
11.02-2 ex	Groats, meal and pellets or barley
*13. 11.08 PL	Starches and insulin
*14. 12.01-2 ex	Peanuts, except for oil extraction
15. 12.08-3-(1)	Edible seaweed, wet or dried, formed into rectangular papery sheets not more than 430 square centimeters each
12.08-3-(2)	Other edible seaweed of specified genera
12.08-3-(3) ex	Other edible seaweed of specified genera
12.08-4	Tubers of konnyaku (devil's tongue), whether or not cut, dried or powdered
*16. 16.02-2 ex	Other prepared or preserved products consisting wholly or chiefly of beef or pork, except ham and bacon
*17. 17.02-1-(2)	Glucose, not containing added sugar or flavoring and coloring material
17.02-3-(2) ex	Lactose, not containing added sugar or flavoring and coloring material, less than 90 percent milk sugar by weight

Table 12.8 (Continued)

	17.02-5 L	Sugar syrup, except maple sugar
	17.02-6 L	Caramel
	17.02-7 L	Artificial honey
	17.02-8-(1/2)-B L	Other sugars and syrups, except sorbose and hi-test molasses
*18.	20.05 ex, L	Fruit purees and fruit pastes made from citrus fruits (except lemons and limes), pineapples, grapes, apples or peaches
*19.	20.06-1-(1) L	Pineapples, otherwise prepared or preserved, containing added sugar
	20.06-1-(2) ex, L	Fruit pulp made from citrus fruits (except lemons and limes), grapes, apples or peaches
	20.06-2-(1) L	Pineapples, otherwise prepared or preserved, not containing added sugar
	20.06-2-(2) ex, L	Fruit pulp made from citrus fruits (except lemons and limes), grapes, apples or peaches, not containing added sugar
*20.	20.07 ex	Fruit juices except made from lemons, limes, grapefruit, cherries, apricots, berries (excluding blueberries and strawberries), prunes, mangos and other tropical fruits (excluding pineapples) and sloe bases
	20.07-2 ex, L	Tomato juices with a dry weight content less than 7 percent
*21.	21.04-1-(1) L	Tomato ketchup and tomato sauce
*22.	21.07-2 ex	Ice cream powder, prepared milk powder for infants and other preparations consisting mainly of milk; seaweed preparations; mochi, cooked rice, roasted rice flour, mijinko, kanbaiko, vitamin-enriched rice and similar preparations of rice, wheat and barley; and other food preparations (except rations, peanut

Table 12.8 (Continued)

> butter, sweet corn, Korean ginseng tea and those with less than a 50 percent sugar content)

Note: Starred items represent the twelve groups of products included in the United States initiated GATT complaint. Ten of these were found by GATT to be in violation of Article 11 in 1987. Japan accepted the results in 1988. An "ex" after a CCCN number indicates that only the specified items are subject to quota. If no "ex" appears, all items falling within the CCCN number are under quota. An "L" or "PL" next to the CCCN number indicates that the product was found to be in violation of Article 11 and with the former being gradually "liberalized" by April 1990 and the latter "partially liberalized" by the same date.

Source: MacKnight 1987, 1989; Zenkoku Nōgyō Kyōdō Kumiai Chuokai 1987(b).

items (most notably rice), are authorized by GATT as state traded rather than privately traded. Those items protected through state trading in Japan are condensed milk, dried milk, butter, milk powder, whey powder, wheat, meslin (a mixture of wheat and rye), barley, rice, raw silk, tobacco, salt, alcohol, and opium (Yoshioka 1982:349). Table 12.9 lists items restricted by the United States and the European Community in 1980.

FOOD SECURITY: WHEN DOES FREE TRADE TURN INTO FOOD CONTROL?

Japan's bargaining stance at GATT maintains that rice is tied to the issue of food security, and is a government traded commodity that is not covered by Article 11. It is true that Japan did not dump its surplus rice on the market during the rice gluts of the late 1960s and late 1970s. Instead it chose to stockpile the old rice and internally regulate rice production. Many Japanese still identify with periods of food shortage. It is not surprising then that a 1987 survey conducted by the Prime Minister's Office found that over 70 percent of Japanese support the idea of self-sufficiency (Furuzawa 1988:91). Many parts of Japan experienced famines during the last century and still have vivid memories of postwar food shortages.

Table 12.9
Residual Agricultural Import Restrictions in the United States and
the European Community

Country	Number of Items	Major Items Restricted
United States	1	Beet sugar, cane sugar[a]
Canada[b]	4	Milk and cream, butter, cheese, curd
W.Germany	3	Potatoes, potato starch, prepared vegetables
United Kingdom	1	Bananas
Italy	3	Bananas, grapes, citrus fruit juice
France	19	Horse, sheep, mutton, natural honey, cut flowers, vegetables, bananas, pineapples, grapes, prunes, melons, prepared fish products, etc.
Benelux	4	Cut flowers, vegetables, grapes, etc.

[a] The United States has additional import restrictions under GATT. These include waivers on 13 items BTN (1-24 group) including wheat, flour, processed wheat products, peanuts, milk and cream, butter, and dairy products. The importation of beef is also regulated under a different law.
[b] EC farm product imports receive surcharge levies to maintain common support prices within the EC region. These are (1) basic type surcharge--pork, eggs, poultry meat; (2) surcharge for processed products--pork, eggs, poultry meat ; (3) surcharge plus tariffs--beef; (4) surcharge plus deficiency payment--durum, olive oil; (5) tariffs plus adjustment levies--vegetables, wine, etc.

Source: Nōrinsuisanshō 1979:61.

Likewise, there is keen awareness that Japan is a resource poor country and that this led to Japan's role in World War II. For this reason there is also concern regarding a secure supply of oil from the Middle East because without it, the tractors and rototillers would not have fuel.

Table 12.10
Changes in the Self-Sufficiency Rate for Consumable Agricultural Products in Japan 1960-1985

	1960	1965	1970	1975	1980	1982	1983	1984	1985
Overall Self-Sufficiency Rate	91	83	79	76	70	71	70	72	73
Grains	83	62	48	44	29	31	30	34	34
Rice	102	95	106	110	87	93	94	109	107
Wheat	39	28	9	4	10	12	11	12	14
Oats and Barley	107	73	34	10	15	16	15	16	15
Beans	44	25	13	9	7	9	7	9	8
Soybeans	28	11	4	4	4	5	4	5	5
Vegetables	100	100	99	99	97	96	96	95	95
Fruit	100	90	84	84	81	79	81	74	76
Eggs	101	100	97	97	98	98	98	99	98
Milk and Dairy Products	89	86	89	82	86	85	86	86	89
Meat	91	90	89	77	81	80	80	80	81
Beef	96	95	90	81	72	71	70	72	72
Pork	96	100	98	86	87	87	85	84	86
Sugar	18	30	23	16	29	31	30	32	34

Source: Nihon Nōgyō Nenkan Kankōkai 1988:523.

Overall grain self-sufficiency dropped from 83 percent in 1960 to 34 percent in 1985 as shown in Table 12.10. The overall food self-sufficiency rating was 73 percent in 1985, well behind other industrialized countries.

Japan's caloric self-sufficiency was 52 percent in 1982 while the United States, Great Britain, France, West Germany, Italy, and Holland had rates of 150, 70, 121, 77, 79, and 86 respectively

(Norintōkei Kyōkai 1986(b):100). Among the EC countries in 1982 only Holland scored lower than Japan in grain self-sufficiency.

From 1960 Japan has steadily liberalized imports of agricultural products. Japan liberalized 121 agricultural and fishery products in October 1960; soybean imports in July 1961; eggs in October 1962; bananas, sugar, and twenty-three other items in April 1963; grapefruit and nineteen other items in June 1971; and pork and pork products in September 1971. As a result agricultural trade, especially with the United States, has risen significantly and represented about one-fifth of all United States exports to Japan in 1988.

Table 12.11 shows the countries which sell agricultural products to Japan. The United States, by far the largest exporter of agricultural products to Japan, provides 33 percent of Japan's food imports. The leading items are soybeans, corn, wheat, sorghum, and beef.

There are, however, several reasons for Japan to be reserved about increasing its reliance on the United States. For example, during the "Soybean Embargo" of 1973 President Nixon briefly banned soybean exports to prevent a (mistaken) shortage predicted to cause inflation. This is called the "Soybean Shock" in Japan because it had to quickly locate soybeans for tofu and livestock feed.

Some fear that United States rice imports would be unreliable due to fluctuating prices and inadequate water supply, possibly leading to a situation similar to the Soybean Shock of 1973. Water supply is particularly worrisome because there are only about 250-350 millimeters annual precipitation in northern California. Water is stored in reservoirs to be shared by farms and nearby cities such as San Francisco. In 1976 Koda Farms, famous for the Kokuhō Rose variety of California, produced one-third of its projected rice output due to a water shortage.

Further, much of United States food policy is tied to foreign aid as a political tool. The Japanese realize that the United States essentially forced the sale of surplus rice to neighboring Korea. The United States has also used food, as in the case with sugar and Cuba, as a political tool.

Table 12.11
Country of Origin for Major Agricultural Imports to
Japan in 1987 (in millions)

	Quantity	Dollar Amount	Share
Total Agricultural Products:			
All Countries	---	18,046.7	100.0
United States	---	5,995.4	33.2
Australia	---	1,556.8	8.6
Canada	---	991.5	5.5
Thailand	---	706.9	3.9
China	---	1,565.6	8.7
Taiwan	---	979.2	5.4
Philippines	---	448.3	2.5
South Africa	---	295.9	1.6
New Zealand	---	469.3	2.6
Brazil	---	472.0	2.6
Soybeans:			
All Countries	4,817	1,072.4	100.0
United States	4,332	958.8	89.4
China	323	76.3	7.1
Brazil	128	28.2	2.6
Canada	26	7.2	0.7
Argentina	---	---	---
Corn:			
All Countries	14,653	1.648.3	100.0
America	9,244	1,045.8	63.4
South Africa	1,280	142.2	8.6
Thailand	60	7.1	0.4
Wheat:			
All Countries	5,620	885.5	100.0
United States	3,241	489.0	55.2
Canada	1,377	253.3	28.6
Australia	1,002	143.2	16.2
Raw Sugar:			
All Countries	1,813	276.3	100.0
Australia	470	68.9	24.9
South Africa	332	50.7	18.3
Philippines	---	---	---
Cuba	577	81.3	29.4
Taiwan	43	6.4	2.3

(continued)

Table 12.11 (Continued)

Thailand	391	69.0	25.0
Grain Sorghum:			
All Countries	4,976	496.3	100.0
United States	2,079	208.8	42.1
Australia	748	80.9	16.3
Argentina	1,578	151.9	30.6
Barley:			
All Countries	1,363	145.8	100.0
Canada	772	79.8	54.7
Australia	529	59.1	40.5
United States	61	6.7	4.6
Beef:			
All Countries	179.1	553.5	100.0
Australia	105.2	281.5	50.9
United States	63.4	238.9	43.2
New Zealand	6.0	18.7	3.4
Pork:			
All Countries	207.7	1,040.6	100.0
United States	14.7	76.9	7.4
Canada	22.4	112.1	10.8
Denmark	78.1	391.9	37.7
Taiwan	83.9	417.0	40.1
Sweden	2.8	13.7	1.3
Bananas:			
All Countries	764.6	377.6	100.0
Philippines	620.5	308.1	81.6
Taiwan	82.4	44.3	11.7
Ecuador	57.0	23.5	6.2
Coffee Beans:			
All Countries	242.5	1,015.1	100.0
Brazil	48.6	234.7	23.1
Columbia	40.1	180.5	17.8
Indonesia	49.5	153.0	15.1
Uganda	5.4	18.3	1.8
Guatemala	8.6	40.4	4.0
Honduras	19.1	88.3	8.7

Source: Nihon Nōgyō Nenkan Kankōkai 1988:511.

JAPANESE GOVERNMENT PLANS FOR
AGRICULTURE IN THE TWENTY-FIRST CENTURY

The Ministry of Agriculture Planning Reports

In 1986 the Agricultural Policy Deliberation Council (APDC) published its directive for the future in a publication titled "The Basic Direction of Agricultural Policy Aiming at the Twenty-first Century" (see Appendix A). This planning directive was a modification of the earlier plan by the committee in 1980 titled "Basic Direction of Agriculture Policies in the 1980s" (see Appendix A) and is the major planning device for the Ministry of Agriculture. In the 1986 report, the APDC argued that while taking into consideration industrial, social, land, and consumer policies along with the ideal of international harmony, the goal was to protect Japanese farming, especially rice production, by bringing domestic prices more in line with international prices. It seems the government has heeded this advice. Subsidies the government pays to rice producers were decreased by 5.9 percent in 1987 and 4.6 percent in 1988.

Although non-rice production was in line with European Community scale and levels, rice production lagged behind international standards. The recommended way to accomplish an independent internationally competitive agricultural sector was to give incentives for creating larger farms. Farm scale was to be enlarged through the transfer of land, in particular land tilling rights, to "core farms" (chūkaku nōka), those farms where there is a male over sixteen and under sixty years of age who works over 160 days a year as a farmer. Efforts are underway to stem the decrease of such farms, which numbered 25.2 percent of the total in 1975 but constituted only 19.8 percent of the total in 1985 (Nōrinsuisanshō Daijin Kanbo Kikakushitsu 1986:137). One of the principle means towards this end is leasing, which was described in the last chapter. The number of organizations, formed by core farmers, which utilize leasing increased twofold from 1975 to 1985, although leased land remained a fraction of the total land.

Under the APDC's policy, prerequisites for an internationally competitive farm system would be large-scale mechanized farming systems with high speed rice transplanters of the four to five row type, along with the increased use of combines. Farm scale would be forty to sixty hectares for large-scale mechanized farms and twelve to twenty-four hectares for medium-scale farms.

Each farm would contain contiguous thirty are (0.72 acres) plots equipped with irrigation and drainage systems to facilitate the rotation of rice, wheat, barley, and soybeans. If such conditions were met, the APDC predicted that by 1995 large-scale mechanized farms could lower the amount of labor required for growing rice to 30 percent of that now required, and lower the cost per unit of product to 40-60 percent of the present figure. Medium-scale mechanized farms could reduce working hours per unit area to 30-40 percent and the cost per unit of product to 50-70 percent of that currently necessary.

In addition, the APDC recommended reducing price subsidies for rice and rice diversion crops. Instead it recommended using the funds to strengthen the agricultural infrastructure, for example, the Land Improvement Project. The APDC also mentioned export of agricultural high-technology products such as biotechnology and recommended that Japan cooperate with GATT in formulating new rules.

The APDC advised that the Japanese maintain a stable food supply and Japanese-style diet (<u>Nihongata shokuseikatsu</u>) which conforms ideally to that espoused by the World Health Organization. Likewise, the vitality of rural society should be preserved by providing opportunities for young people to enter agriculture. Finally, it counseled that terraced rice farming should be allowed to continue, as it preserves the environment by preventing soil erosion and by converting surface water to ground water.

The Maekawa Reports

The status of Japan as an economic superpower contrasts to a standard of living below that of the rest of the industrialized world. In view of this, Prime Minister Nakasone commissioned an advisory panel in 1985 to plan a restructuring of priorities for the future. Formally known as the Advisory Group on Economic Structural Adjustment for International Harmony, the group was headed by Haruo Maekawa, former governor of the Bank of Japan, and called the Maekawa Commission. The group issued reports in 1986 and 1987 outlining methods by which Japan could reduce its huge trade surplus and preserve harmonious relations with the rest of the international community. The main points of the 1986 report concerned the need to boost domestic demand while improving living standards, restructure domestic industry to meet domestic rather than external demand, expand imports of manufactured goods by reducing tariffs, increase direct investment and foreign aid abroad, and reduce the national budget deficit.

The second report made new suggestions. The commission advised the government to increase its support to and tax breaks for home buyers, eliminate the tax disparity between urban agricultural and residential land[3], promote domestic consumption by raising wages and ensuring that consumers receive their share of the benefits of yen appreciation. It also recommended that the work week be shortened and leisure time expanded, industrial sophistication be enhanced and employment opportunities expanded with forward-looking plans rather than focus attention on the "hollowing out" of Japan. In addition, it counseled that domestic market access to foreigners be improved, imports be increased, and allocation of public works projects to depressed regions be a priority.

The Maekawa report echoed the APDC report regarding agricultural policy. It mentioned the necessity of improving productivity, enlarging the scale of rice production, and developing biotechnology. It also supported maintenance of self-sufficiency in rice production but felt that the amount of independently routed rice should increase (thereby reducing the amount of subsidies), and advised that the government introduce competitive situations in each stage of marketing the rice. For non-rice agricultural products, it advised reduction of national border adjustment measures to a minimum (Keizai Kikakuchō Sōgō Keikakukyoku 1987).

Economic Planning Office Estimates

Current forecasts by the Economic Planning Office predict that the number of core farmers will actually decrease from 20 to 14 percent of all farm households between 1985 and the year 2000, although the amount of acreage devoted to rice will increase. Acreage used for rice by full-time farmers would increase from 1.2 hectares to 2.5 hectares while rice acreage for part-time farmers would decrease slightly (Keizai Kikakuchō Sōgō Keikakukyoku 1987: 38).

The Fourth National Comprehensive Development Plan of 1987

Every ten years the National Land Agency publishes a comprehensive development plan for Japan. The fourth such plan, called Yonzenso, recommended "A basic direction of regional activity and stable food supply" (Kokudocho 1987:65). Preserving the natural environment and national land were given top priority followed by the enlargement of the scale of production, raised productivity, and introduction of new technology. The plan also outlined steps to facilitate marketing of agricultural products from distant rural areas. Social factors such as the rapidly aging rural population in Hokkaido, northern Tohoku, and Southern Kyushu, and the fostering of young core farmers with an enlarged scale of production, were also addressed.

AN ECOLOGICAL APPROACH TO JAPANESE AGRICULTURE: PEOPLE, LAND, AND THE ENVIRONMENT

People

The main purpose of this book has been to examine the interrelationship between people, their land, and restructuring of their environment. To the average farm household in Nakada Township, there can be no cost reduction in production or enlargement of scale unless their livelihood is secure. As Chapter 7 has shown, farmers are willing to lavish money on new machinery to save time which can be better spent earning wages in factories, and eagerly take risks if it is clear that it will improve their lives and that there will be long-term benefit. The central components of any solution to the farm dilemma in Japan must involve the cultural right of the members of the farm household to perpetuate their corporate body both through succession, inheritance of land, and ancestor worship.

Having conducted fieldwork in rural Japan for almost two years, I conclude that the main problem encountered by rural society is household succession and inheritance. As single heir to the household, rural farmers bear the responsibility for keeping up the ancestral graves, the household Buddhist altar, coordinating the anniversary Buddhist memorial services, sustaining hierarchical or reciprocal social relations with branch or main households in their ancestral line, and finally to serve as an important link between rural and urban relatives who at some time in the past migrated to the city as non-heirs. The annual gift rice (zotōmai) sent to urban relatives is a symbolic gesture of this reciprocal relationship and mutual dependence. At the summer Obon Festival, it is not just the ancestral spirits who return home to the rural areas. Urbanites, eager to renew their own ancestral identity, jam the trains to the distant yet spiritually close "hometown" (furusato).

The bride shortage and aging rural population are evidence of succession and inheritance problems. Women were driven out of the paddies through the introduction of the rice transplanter during the 1960s and now they will be the first to be displaced again through "hollowing out." The possibility that women will be the first to lose their jobs due to hollowing out as explained in Chapter 9 can only exacerbate the exodus of women to the cities.

The average rural age has been rising not just because of the rise in Japanese longevity. It has also been rising because some households do not have an heir. Such households are more likely to keep their land and lease it rather than sell it and the problem of infirm elderly without someone to take care of them seems to be on the rise.[4] In households who do have an heir, the problem of infirm in-laws will continue to be a burden on farm wives unless improved social services are provided.

Another growing "problem" for the hamlet is how to promote harmonious relationships between households. Marginalization of the landless and growing disparities between landed, stemming not only from the amount of land owned, but also from the relative difference in job security as documented in Chapter 9, may be the central problem for farming communities of the twenty-first century. This disparity may lead to impermanence and instability in rural areas and create an imbalance in the industrial work force which currently is utilizing the stability of a work force tied to the land.

Likewise, the income parity between rural and urban households may also disappear if a safety net is not created to save those households which are affected by the loss of subcontracting jobs and cuts in the rice price and rice diversion subsidies. At present these policies support the rural farmers in a way comparable to France's idea of "retiring the peasants." Without the subsidies, a rural-urban income disparity would lead the way to a collapse of the rural economy in Japan.

Land

Unless the government secures the farmers livelihood, they will not sell or lease (so-called "entrusting (juitaku)" their land. Most farmers grow rice in their own name by contracting out (itaku) parts of the production process such as custom tilling or transplanting. They are afraid that leasing rights (jutaku kenri) as a form of land tillage rights (kosaku kenri) might at some future date be redefined as a type of tenant right (kosaku kenri). This is possible because government promoted "core farmers" would be dependent upon lease arrangements and paddy consolidation of at least several small-scale farms. About one-third of the farming households in 1989 contracted out work rather than risk any transfer of land rights that may evolve out of a leasing situation. Others participate in unregistered leasing, keeping the paddy in their own name in the tillers register at the Land Commission and Land Improvement District offices. Thus, as Toshihiko Isobe (1988) suggests, government plans do not address key problems of farm households. All the government plans for the future that are listed above mistakenly assume that they can correct the problem of scale by making policies based on a division of family farms into small, medium, and large according to acreage under cultivation. This ignores the important elements of kinship, land ownership, land use rights, and the complex social relations upon which farming is dependent. Accordingly, in a survey of farm successors taken in 1987, 81 percent of those interviewed answered "yes" to the question "Do you think that it is essential to consider agricultural production at the hamlet level?" (Miyagi-ken Nōgyō Kaigi 1987:64).

In planning for the future, Nakada farmers cannot help but consider the fate of the Tadano farm and the Hachirōgata Land Reclamation Project. The Tadano farm, located in Nakada Township, was at one time the largest privately owned farm in Japan. It was split up upon the death of its founder. In the case of the Hachirōgata Reclamation Project, three Nakada Township households sold their farms in the late 1960s to participate in this large-scale government planned rice growing project. Now the project, once famous as a "pioneering" large-scale project, is well known in Japan for its open resistance to government policies that limit large-scale production of rice. Farmers there sell black market rice despite police efforts to stop the practice.

For these reasons farmers in Nakada Township share reservations with other Japanese farmers concerning enlargement of farm scale with other Japanese farmers. In the traditional village where I conducted research in 1983, 807 households were surveyed by the township regarding their attitudes towards enlargement of farm scale. Approximately 64 percent desired their farm size to remain the same, 20 percent wanted enlargement, 5 percent wanted to decrease farm size, 9 percent were uncertain, and only 1 percent wanted to quit farming.

Japanese farmers realize that rice production on average Japanese three acre farms cannot compete with the 620 acre average farm size in California. The government planned scale increases may well make Japanese farmers competitive in the world rice market. To increase the scale of their farms to any meaningful extent by purchasing land is not possible because a one-quarter acre paddy may cost $15,000 to $25,000 depending on its location. The only way to increase scale would be to lease land domestically or buy it in California, which is certainly a possibility, unpopular as it would be to Americans. No viable alternatives guaranteeing the household's future have yet been presented, so farmers understandably have misgivings about increasing their farm scale. Tables 12.12 and 12.13 illustrate this pessimism. Farmers fear that the government will lower the producer's rice price and that they will be forced to lease out their land.

Yoshikazu Kanō (1987) and A.B.A.R.E. (1988) have suggested allowing free market forces to take over. Kanō argues for a two-step reform of the food control system and eventual liberalization of foreign rice. The first step would be to make the idle crop plan optional by attaching it to a price stabilization program. Farmers could choose whether or not to idle their land and receive government subsidies below market price. The efficiency, cost effectiveness, quality, and yield of full-time farms which survive would be enhanced through the competition. These survivors would be ready for the second step, international free market competition[5].

Others, such as Kenichi Omae have argued for free trade on rice as well as incentives for farmers to sell their land. He complains that both the price of rice and of land is much too high from the consumers point of view and that a deregulation of rice would put pressure on Japanese farmers to sell their land. This, it is hypothesized, would alleviate the pressure on land prices and make the purchase of urban housing much cheaper.

276

Table 12.12
Public Opinion Survey Results Regarding the Producer's Rice Price

Question: If the domestic market were liberalized, what do you think would happen to the producer's rice price?

Prefecture	Less than 10,000 Yen	About 13,000 Yen	About 15,000 Yen	A Little Lower Than Now	About the Same	A Little Higher	Don't Know and Other
Fukushima Prefecture	8.2	20.0	14.5	30.1	19.1	5.5	2.7
Miyagi Prefecture (North of Sendai)	4.9	13.2	16.7	23.6	25.0	9.0	7.6
Niigata Prefecture (a)	5.1	21.5	----	50.6	15.2	5.1	2.5
Niigata Prefecture (b)	10.5	10.5	13.7	24.2	23.4	8.9	8.9
Mie Prefecture	7.3	30.9	23.6	18.2	5.5	10.9	3.6
Nagano Prefecture	3.8	23.1	7.7	19.2	19.3	13.5	13.5

Source: Kanō 1987:52.

Table 12.13
<u>Public Opinion Survey Results Regarding the Future Number of Japanese Farms</u>

<u>Question</u>: What will have happened to the number of rice producing households in your region ten years from now?

<u>Prefecture</u>	<u>Same as Now</u>	<u>30 Percent Decrease</u>	<u>Half</u>	<u>Lower Than One-Third</u>	<u>No Response</u>
Fukushima	36.4	29.1	20.9	13.6	0.0
(Northern) Miyagi	29.9	46.5	13.2	9.0	1.4
(Southern) Niigata	22.8	54.4	16.5	2.5	3.8
(Middle) Niigata	25.8	49.2	16.9	7.3	0.8
Mie	18.2	52.7	18.2	7.3	3.6
Nagano	15.4	36.5	17.3	25.0	5.8

Source: Kanō 1987:52.

In 1988 the price of land in Osaka increased 51 percent, Kyoto 93 percent, and Kobe 96 percent. According to the Economic Planning Agency, land in Japan costs one hundred times that of land in the United States. This means that housing is beyond the reach of most urban middle class Japanese who have not inherited land from the previous generation. High inheritance taxes coupled with the need for large downpayments on home loans make it difficult even for those who have inherited financial resources.

Theories such as Omae's are mistaken since farmers would not necessarily sell their land. Even if they quit growing rice, most of their income (about 85 percent) comes from working side jobs. Even Yasuhiko Yuize's (1987) suggestion that property taxes on small-scale farms be raised would probably not be enough to force households to sell land, especially if it continues to appreciate.

Regardless of the price of foreign rice, Japanese society will continue to place a high premium on the historical and cultural significance of rice. It is difficult to find an urban Japanese in Tōhoku cities who has not received a gift from relatives of the fresh crop of rice (shinmai). Thus, the production cost comparisons given in Table 12.3 mean less because they are based on one sector only. They also do not consider efficiency from the point of view of the household, which, in order to perpetuate itself requires members to be employed in both agriculture and wage labor. Single sector rationality, such as viewing costs and benefits from either the industrial or agricultural sector, leads to the mistaken conclusion that Japanese industry is more efficient that Japanese agriculture. To the contrary, using such irrational logic, one might suggest that Japanese industry is efficient because agriculture is not.

The essential factor is cultural. How can households and communities maintain a survival strategy in line with their prescribed value system? The answer is not in a single factor economic rationality but rather in the articulation of production with local history; kinship, residence, and inheritance[4] rules; religion; politics; and ecology.

Environment

As Tsutomu Ōuchi (1989:65) points out, the productivity of agriculture and forestry is dependent on environmental factors such as the climate, the topography, soil quality, and the amount of rainfall. This is unlike manufacturing and the service sector. Both Ōuchi, Nobutane Kiuchi (1988), Toshihiko Isobe (1988), and Yukio Honda (1982) have emphasized the valuable role that rice production plays in water and soil conservation in Japan, an archipelago with little land that is flat enough for large-scale farming. Both Ōuchi and Kiuchi argue for a flexible situational approach to international trade as opposed to the United States position which might be termed absolutist or at least fixed in "principle."

As Kiuchi states, it may be possible for Japan to be a leader in re-orienting GATT rules towards a more ecological viewpoint. Environmental adaptability (shizen junnōsei) and cultural adaptability (<u>bunkateki junnōsei</u>) should be the standards by which agricultural products are traded on the world market. By environmental and cultural adaptability I am referring to the interrelationship between people and their environment.

The criteria for environmental adaptability might include consideration for the preservation of the natural environment through analysis of the composition of the biological community including the species, numbers, biomass, and distribution in time and space. The quantity and distribution of non-living material such as sunlight, temperature, water, and minerals would need to be considered. Likewise, rates of energy flow through the ecosystem and rate of nutrient cycling would necessarily be measured. Finally, worldwide standards for ecological regulation need be developed. These would include ideas and means for the world population to have a balanced relationship with its environment.

Japan and the rest of the world will need to evaluate the degree to which countries can exploit resources and alter non-domestic ecosystems outside their territorial limits while claiming to preserve the natural environment at home. At present land-limited Japan utilizes "ghost acres" (Borgstrom 1967) in the world seas and foreign countries in order to maintain a high population density in a small area. Japan's low self-sufficiency rating means that a resource base outside of Japan ("ghost acres") are used to maintain Japan's domestic economy. The resulting importation of food from abroad, over-fishing of the global sea commons, and funding of environmental destruction in the world commons or in the territory of other nations weaken Japan's claim to protection of its domestic ecological integrity.

Finding a balance between environmental preservation and culturally prescribed material luxuries will be the central problem of the twenty-first century. If Japan chooses to argue along the lines of cultural adaptability, it should adopt a vigorous foreign policy that would further each society's right to healthy food, locally grown fresh food varieties which are suited to specific climate and soil conditions and one which would conserve finite natural resources while promoting food security and indigenous kin rules necessary for the preservation of family farms. If this was done Japanese farmers might find a new role for themselves within the framework of internationalization (<u>kokusaika</u>) and an information based society (<u>jōhōka shakai</u>), two dominant trends Japan seems to be pursuing for the 1990s.

Agriculture is more than the production of cheap food. New quality standards for GATT agricultural trade products would provide farmers worldwide with an opportunity for grass roots organizing and international exchange of agricultural information. Likewise, it would present cultural criteria for a much needed reorganization of Japanese and world agriculture. Such environmental and cultural criteria would recognize the need to promote and protect those rights which originate from indigenous history, philosophy of life, traditions, social structure and ecology.

Recent environmental reports have stressed the rapid depletion of the world's natural resources and growing desertification and erosion. It may be worthwhile for Japan and the United States to reconsider the words of F. H. King in his <u>Farmers of Forty Centuries</u> (1911:113) when he first visited Japan:

> ...Such enormous field erosion as is tolerated at the present time in our southern and south Atlantic states is permitted nowhere in the Far East, so far as we observed, not even where the topography is much steeper...The practice (of composting and terracing on mountain slopes), therefore, gives at once a good fertilizing, the highest conservation and utilization of rainfall, and a complete protection against soil erosion. It is a <u>multum in parvo</u> treatment which characterizes so many of the practices of these people, which have crystallized from twenty centuries of ... experience.

Because family farming links the temporal continuity of <u>ie</u> members, both dead and living, with the spatial integrity of farmland, the family farm has remained the backbone of rural Japan. Likewise, the small-scale of agriculture links people with the natural environment in a way which has prevented rapid degradation of the Japanese archipelago. The real issue of Japanese agricultural development is not production cost comparisons but rather whether holistic cultural values stressing long-term cultural adaptation can mitigate concerns for short-term profit domestically and internationally. In light of the growing concern over international environmental issues, Japanese farmers and policy makers have an unprecedented opportunity and responsibility to redefine their own identity and teach their own children, fellow citizens, and people of the world about high quality standards regarding the use of space and time in local environmental situations.

NOTES

1. Keidanren, the Federation of Economic Organizations, is the leading private nonprofit economic organization representing all branches of economic and industrial activity in Japan. As of September 1986, Keidanren's membership comprised 121 industrial federations and regional economic organizations, 877 corporations, and 31 other organizations. The Japan Federation of Employers' Association, known as Nikkeiren, has also called for a gradual decontrol of the rice industry.

2. In the twenty-three Tokyo wards, the tax rate on residential land was eighty times that of agricultural land.

3. Financially the elderly are well taken care of, being covered by the Farmers Pension Fund Act of 1970 (see Appendix A), the National Pension Scheme as reformed in 1988, and the Employees Pension Insurance Scheme for wage earners.

4. This has also been suggested by Takekazu Ogura (1982:614) in his concept of "viable farms."

Appendix A:
Major Agricultural Policies
from 1968 to the Present

1873 MEIJI LAND TAX
(<u>Chiso Kaisei Jōrei</u> 地租改正条例)
Tax on privately owned land was fixed at 3% of the land value
and was paid in cash. The land value was based on the quality
and fertility of the land and land map registers were sketched.
This cash tax based on land value changed the <u>kokudaka</u> system
of the Tokugawa government whereby rice was paid to the
feudal lord in kind based on the productive yield of the land.
Ownership of land was changed from the shōgun, local feudal
lords, their bannermen and retainers to a non-feudal private
ownership system. A very low tax rate for previously untaxed
grass and forest lands was established. One year before this
the <u>nago-hikan</u> indentured servitude system was also abolished.

1876 FOREST AND GRASSLAND CLASSIFICATION SYSTEM
(<u>Nōrin Genya tō Kanminyū Kubetsu Seido</u> 農林原野 等
官民有区別制度)
Forest lands were divided into government owned <u>kanyūchi</u> and
municipally owned <u>minyūchi</u>. Forest lands used as commons
(<u>iriaiken rinya</u>) were a separate category of municipally
(village) owned land. In 1966 commons became legally subject
to private ownership through the Common Forest and Grassland
Modernization Law.

1890 IRRIGATION COOPERATIVE LAW (revised in 1902)
Irrigation cooperatives were legally established for landowners.
See the 1949 Land Improvement Law.

1898 MEIJI CIVIL CODE
(Meiji Minpō 明治民法)

This code legalized the traditional samurai class family system. Households were registered under the 1872 Household Registration Law that recorded the social status of the household, rank, birth and death dates of all individuals who egressed or ingressed to the local district, and the rank and parentage of all individuals of the household. The institution of household head (koshu) was legalized and reinforced by the provision that at least one-half of the property, the "legally secured portion," accompany household succession. The right to succession fell to first born males although certain "imperative reasons" were loosely interpreted in order to disqualify the eldest son and enable others to succeed. Unborn children and adoptive children had rights to compensation for damages and a right to succession. A woman could succeed when the household head died without a male heir although she would have to transfer the position to a male in the event she married or adopted a son. Direct lineal descendants of an ancestor had the first right to inherit property even over the spouse or household head in the event that the latter was not the most direct lineal descendant. Persons of equal lineal distance were to receive equal shares.

Private and group usufruct rights over land and houses were not mentioned. However, the commons right (iriaiken) was guaranteed as a type of joint ownership. Several types of tenancy were defined including permanent tenancy. Emphyteusis (eikosakuken), whereby the tenant right could be inherited or leased out, was redefined to exist when a person tenanted land more than twenty years. However, the practice was limited to less than fifty years in the attempt to fix private property rights for taxation purposes. This limit, however, was not systematically enforced. See 1948 Civil Code.

1899 ARABLE LAND REORGANIZATION ACT
(Kōchi Seiri Hō 耕地整理法)

Agricultural plots were formed into 1 tan (2.47 acre) rectangular plots and consolidation of fragmented land holdings was encouraged. The project was directed at rice paddies rather than field crop land. Six times more paddies than field crop land were reorganized by 1926. The act formed the legal basis for the 1909 Land Reorganization Projects which were undertaken by Arable Land Reorganization Associations (Kōchi Seiri Kumiai). The membership was comprised of both landlords and cultivating tenants while the leadership was restricted to landlords. This law was abrogated in 1949 with the establishment of the Land Improvement Law.

1899 AGRICULTURAL ASSOCIATION LAW
(Nōkaihō 農会法)
Agricultural cooperatives started through the unpassed 1891 Credit Cooperatives Bill were subsidized. Cooperatives were established with the voluntary membership of farmers who had paid at least two yen of land tax or landlords possessing over four tan (forty ares) of land. The associations were mainly involved with the improvement of agricultural techniques and mediation in the marketing of vegetables. The associations were prohibited from commercial buying and selling, which was done by the Industrial Cooperative Associations. The association was transformed during World War Two into a wartime organization for agricultural policy control and finally merged with the Industrial Cooperative Associations in 1943.

1900 INDUSTRIAL COOPERATIVE ASSOCIATION LAW
(Sangyō Kumiai Hō 産業組合法)
Credit, purchasing, production, utilization, and marketing agricultural cooperatives were established in rural areas. These rural agricultural cooperatives were subdivided into types according to limited, unlimited, and guaranteed liability. During the first few years of its development, the number of credit and marketing cooperatives outnumbered production or purchasing cooperatives. The marketing of rice, wheat, barley, and cocoons were the dominant marketing activity. By 1942 guaranteed liability credit cooperatives were predominant. This association is the predecessor of Nōkyō, the agricultural cooperative.

1908 REGULAR WATER USE COOPERATIVE LAW
(Futsū Suiri Kumiai 普通水利組合)
This law established cooperatives for irrigation facility maintenance. These irrigation cooperatives were based on traditional water use rights.

1913 RICE TARIFF
(<u>Beikoku Kanzei</u> 米穀関税)
A duty of 1 yen per 60 kilograms of rice was established for imported rice. Taiwan and Korean (colonial) rice was free from the duty and comprised 5% of the total Japanese rice in 1915. This percentage rose to 30% by 1935 and presented problems for the over-priced domestic rice, which became protected through the 1942 Food Control Act.

1921 RICE LAW
(<u>Beikoku Hō</u> 米穀法)
This law enabled the government to control rice prices by buying, storing, and selling rice as needed. It also gave the government the right to regulate imports.

1923 THE AGRICULTURAL CENTRAL BANK LAW
(<u>Nōrin Chūōkinko Hō</u> 農林中央金庫法)
This bank was created as the central credit organ for agricultural and other cooperatives. Its capital was subscribed mainly by the agricultural cooperatives and their federations. Payments for rice purchased under the Food Control Act were channeled through this bank.

1924 TENANCY CONCILIATION LAW
(<u>Kosaku Chōtei Hō</u> 小作調停法)
Conciliation Unions were established for resolution of disputes between landlords and tenants. The number of unions grew steadily until 1940.

1925 ELECTION LAW REVISION
(<u>Senkyo Hō Kaisei</u> 選挙法改正)
Universal male suffrage for males over 25 years of age was achieved when male tenants received the right to vote. The age was lowered to 20 when women received the right to vote in 1947.

1926 REGULATIONS FOR ESTABLISHING SMALL OWNER-OPERATED FARMS
(<u>Jisakunō Sōsetsu Iji Hojo Kisoku</u> 自作農創設維持補助規則)
With the consent of the landlord, tenants could take out 25 year loans at 3.5% interest to purchase the land they worked. The price was determined at 16 times the value of the annual rent paid in kind. The program floundered in 1930 when land

values dropped and mortgage subsidy funds from the state stopped. If this policy had been successful in full, it would have only altered 4% of the ownership of tenanted land over a period of 25 years.

1932 FARM RESCUE DIET
(Kyuno Gikai 救農議会)

Two hundred million yen were allocated for employment on public works projects in 12,000 villages. The emperor also announced a gift of 5 million yen for medical aid in the countryside. This established the government slogan "jiriki kosei" (rehabilitation through self-help).

1932-35 RURAL REHABILITATION PROGRAM
(Noson Kyusaisaku 農村救済策)

Each year 1000 villages were chosen to receive money to consolidate debts, extend loans through producer co-operatives, assist village handicraft industries and provide technical aid to farmers.

1933 DEBT CLEARANCE UNIONS ACT
(Noson Fusai Seiri Kumiai Ho 農村負債整理組合法)

Debtors were encouraged to form unions which could negotiate with creditors for easing the debt repayment terms. It also provided limited funds for the conversion of debts into government loans at a uniform 6% interest. The amount paid out, however, was only 0.6% of the total indebtedness. The average small-scale farmer owed about one year's profit.

1938 FARMLAND ADJUSTMENT ACT
(Nochi Chosei Ho 農地調整法)

Forty million yen per year were allocated to loans for creating and maintaining owner-run farms. The goal was 15% of the land under tenancy (about 46% of the total farm land was under tenancy at the time) to be converted to ownership by a million new owner-cultivating farmers (jisakuno). Under this act, the government defined lease conditions more precisely, controlled land prices, and improved the terms for mediation of tenancy disputes while protecting the tenants from eviction.

1942 FOOD CONTROL ACT
(Shokuryō Kanri Hō 食糧管理法)
At this time the government started buying and selling rice, a practice which has continued. In 1979 the government sold rice to retailers at a 12.3% loss while the consumer rice price rose 3.2%. In the immediate post-war period, this act regulated the storage and distribution of rice and controlled the price of rice and the school lunch program.

1945 FIRST LAND REFORM MEASURE BY THE SUPREME COMMAND OF THE ALLIED POWERS
(Daiichiji Nōchi Kaikaku 第一次農地改革)
The maximum land holding of tenanted land permitted to non-absentee landlords was 5 chō (about 5 hectares). This was changed to 1 chō in the following reform.

1946 SECOND LAND REFORM MEASURE
(Dainiji Nōchi Kaikaku 第二次農地改革)
Land available for purchase consisted of all land owned by absentee landlords (fuzai jinushi) not residing in the village and all tenanted land owned by resident landlords in excess of 4 chō in Hokkaido and an average of 1 chō in the rest of Japan. Each county of each prefecture had its own limits which varied slightly according to local situations. The total holding of leased and cultivated land could not exceed 12 chō in Hokkaidō and an average of 3 chō in the rest of Japan. Land Commissions (Nōgyō Iinkai) were established and were comprised of 5 tenants (those owning less than a third of the land they cultivated), 3 landlords (those cultivating less than a third of the land they owned), and 2 owner farmers (who owned more than a third of the land they cultivated). All rents were to be paid in cash and were to be fixed at rates commensurate with the (low) price of land. The rent was not to exceed 25% of the total value of the crop for that year. Tenancy contracts were to be put in writing and their renewal could not be denied. Forest lands were not included in the reform. New buyers of agricultural land were to have a minimum of 3 tan (30 ares).

1947 AGRICULTURAL COOPERATIVES ACT
(Nōgyō Kyōdō Kumiai Hō 農業協同組合法)
The industrial cooperatives and the agricultural associations were merged into a new cooperative system with banking rights

during the Occupation. These cooperatives were originally merged in 1943 under the Agricultural Organization Law.

1948 CIVIL CODE
(<u>Minpō</u> 民法)

Article 1 provided that equal rights be given to husband and wife and to parents and children. Article 3 abolished the institution of household head provided for by the Meiji Civil Code. Article 4 stated that parental permission is not necessary for adult marriage, divorce, or for adopted <u>yōshi</u> style marriage or divorce. Article 7 abolished the so-called "<u>katoku</u>" single heir inheritance system and Article 8 provided for equal inheritance by siblings. The inheritance system was legally changed to provide for the spouse to receive one-half the inheritance and the remainder to be equally divided among the children. In rural areas this law was circumvented by households which desired to keep the property intact. Multiple heirs were permitted to sign legal release forms giving their share of the inheritance to a designated single heir. See 1898 Meiji Civil Code.

1949 LAND IMPROVEMENT LAW
(<u>Tochi Kairyō Hō</u> 土地改良法)

Land improvement authorities (<u>tochi kairyōku</u>) were established to control irrigation water and carry out the land improvement projects. These local authorities replaced the water rights cooperatives (<u>iriai kumiai</u>) and arable land improvment cooperatives (<u>kōchi seiri kumiai</u>) which were often controlled by landlords. The law provided that the authorities maintain and coordinate land holding consolidation (<u>shūdanka</u>), land improvement, conservation, land reclamation, and protection from calamities. In carrying out the national policy encouraging land improvement projects, the land improvement authority coordinates activity between the agricultural cooperative and the national, prefectural, and township governments.

1950 ACT TO PROMOTE AGRICULTURAL MECHANIZATION
(<u>Nōgyō Kikaika Sokushin Hō</u> 農業機械化促進法)

Provisions were made for the installation, inspection, training in, and funding of agricultural mechanization in order to promote higher efficiency and increased productivity.

1952 AGRICULTURAL LAND ACT
(Nōchihō 農地法)

This major land act established a "land to the tillers" policy. Most of the articles were repealed or amended in 1970 and 1980 revisions of this law. The most important articles were:

Article 1:

Agricultural land was recognized as best suited to be owned by the tiller.

Article 2:

No provisions were made for agricultural production corporations.

Article 3:

Limits were established for rights governing land transfer. The permission of the Land Commission Office acting on behalf of the prefectural governor was needed for leasing land. The upper limit of land holdings was set at 3 hectares (12 hectares in Hokkaidō). The lower limit for land needed before qualifying to be able to acquire land was 30 ares (2 hectares in Hokkaidō).

Articles 4 and 5:

Agriculturally zoned land could not be converted to non-agricultural use.

Articles 6 and 7:

Limits to tenanted land were established. A ceiling of 1 hectare (4 hectares in Hokkaidō) was established for resident landlords. Absentee landlordism was prohibited.

Article 19:

Automatic legal renewal of leasing terms was guaranteed unless the tenant was notified at least 6 months to 1 year prior to the expiration of the contract. Tenant rights to cultivate land were to be inheritable. An upper ceiling for rents was established by the Land Commission. Most contracts were verbal.

Article 20:

Cancellation of leasing was not permitted without the permission of the prefectural governor and the consent of both parties.

Article 21:

Rent was controlled by the Land Commission.

Article 22:

Fixed cash payments (teigaku kinnōsei) were to be made in rent payments. Rent in kind (butsunō) was not permitted.

See Amendments to the Agricultural Land Act of 1952 passed in 1962, 1970, 1975, and 1980.

1961 AGRICULTURAL BASIC LAW
(Nōgyō Kihon Hō 農業基本法)

The main objective of agriculture was established to be as follows: "Agricultural productivity should increase in such a way as to reduce the disparity in productivity between agriculture and other industries. Persons engaged in agriculture should earn higher incomes enabling them to make a living comparable to those engaged in other industries." This law provided for policy measures to meet the objectives as follows:

1. Selective expansion of agricultural production, the raising of productivity, and increasing of gross agricultural output.
2. Improvement of agricultural structure through the enlargement of farm size.
3. Stabilization of the prices of agricultural products and maintenance of agricultural income.
4. Rationalization of marketing and the processing of agricultural products.
5. Training farmers and encouraging their family members to seek employment in other industries.

1962 AMENDMENT TO THE AGRICULTURAL LAND ACT OF 1952 (Land Reform Implementation Act)
(Nōchihō Kaisei 農地法改正)

The 1952 upper limit for land holdings could be exceeded provided that the land was cultivated by self-employed labor. Land use could also be changed with the permission of the prefectural governor. This was accompanied by an amendment to the Agricultural Co-operatives Act of 1947. This amendment included the following provisions:

1. Agricultural Producer Corporations: Farmers or agricultural cooperatives could engage in agriculture with limitations on employing labor power, the amount of land to be leased, and the dividends.
2. The Farming Association Measure
3. The Agricultural Land Trust System

1964-1967 THE KENNEDY ROUND OF THE GENERAL AGREEMENT ON TARIFFS AND TRADE (G.A.T.T.)

(<u>Gatto Bōeki Kōshō Kenedi Raundo</u> ガット貿易ケネディラウンド)
Japan joined GATT in 1955 and during the Kennedy Round held in Geneva signed an international grain agreement with the U.S. and other countries. Tariffs on 90 agricultural items were cut by an average of 53%. The soybean tariff, reduced from 13% to 6%, was the largest item and represented 64 percent of the total amount of the tariff reduction. Kennedy Round agreements affected only 31% of the total U.S.-Japan agricultural trade because feed grains were not included.

1965 REWARD FOR COOPERATION IN THE LAND REFORM ACT

(<u>Nōchi Hōshō Hōan</u> 農地報償法案)
Over 1.7 million landlords who lost land in the land reform were compensated a total of 145.6 billion yen over a 10 year period. This was equal to more than one-half the budget for the first 10 years of the 1966 Land Improvement Plan listed below. The rate of compensation was 20,000 yen for every 10 ares of paddy land and 12,000 yen for field land. Compensation to Hokkaidō landlords was 25% of the rate. For cases of over 1 hectare of land, landlords were compensated at a fixed rate, the total of which was not to exceed 1 million yen. For parcels under 10 ares, 10,000 yen were allocated. All compensation was in addition to the nominal sum received by landlords in the land reform.

1966 THE LAND IMPROVEMENT PLAN

(<u>Tochi Kairyō Keikaku</u> 土地改良計画)
The land improvement plan was designed to increase paddy size, improve irrigation and drainage facilities, and consolidate fragmented land holdings. The plans are listed below:

Item	First Plan	Second Plan	Third Plan
Time Period	1966-1975	1974-1983	1983-1992
Budget (in yen)	26,000	130,000	328,000
Goal	40%	63%	70%

(Note: The budget is in units of 100 million yen and the goal represents a percentage of the total agricultural land in Japan.) Source: Nihon Nōgyō Nenkan 1966-1987.

1970 AMENDMENT TO THE AGRICULTURAL LAND ACT OF 1952

(Nōchihō Kaisei Hō 農地法改正法)

The stated purpose of the land act changed from "land to the tillers" to "land for the purpose of efficient use in agriculture" (kōritsutekina riyō). Further decontrol such as the abolishment of the upper limit on land cultivated by a farmer was instituted. The lower limit necessary to qualify for buying or leasing land, however, was raised from 30 ares to 50 ares. Absentee landlordism was permitted in cases where a household moved away from the village. Upper limits on grass and pasture leased out were abolished. Government permission was not necessary if the leasing period was over 10 years and both parties gave written consent to the contract. At the expiration of the contract either party had the right to refuse renewal of the contract.

The ownership of agricultural land by retired farmers was permitted. The agricultural cooperatives were allowed to engage in agricultural production if so entrusted by their members. Restrictions on agricultural corporations were lifted and members could live outside the township where the corporation was located.

Land transfer rights could change with the permission of the Land Commission. Land transfer agencies (non-profit institutions established by municipalities, etc.) were entitled to buy or sell land as well as to lease or rent it. To increase land transfer through agencies, the government subsidized part of the costs (the interest rate) for the agencies to acquire land. The government also paid 10 years rent in advance on land rented and leased through the agencies. By February 1973, 36 agencies were set up and the amount of agricultural land purchased or sold through these agencies amounted to 10,000 hectares.

This act diminished tenant land rights which were particularly strong after 1920. According to this act, landlords were not required to give "departure from agriculture compensation" (risakuhōshō kin) to tenants when landlords initiated the end to leasing contracts. Land tenanted prior to this was still under the old regulations which gave tenants inheritable rights to the land they cultivated. The value of the kosakuken was often

worth as much as one-half the land value. Rent control was
abolished and standard recommended rent levels instituted.
However, local Land Commissions had no legal redress when
landlords raised rents above the recommended levels.

1970 FARMERS PENSION FUND ACT
(Nōgyōsha Nenkin Seido　農業者年金制度)

Administered by the Land Commission (Zenkoku Nōgyō Kaigisho
1982a, 1982b), this act established a retirement system for
farmers. Pensions were received from ages 60 to 65 after
which the National Retirement System (Kokumin Nenkin Seido)
complemented it. The actual amount to be received before the
national retirement system payments would start was about 81%
of the amount to be received after age 65 when the recipient
would receive from both systems.

Besides making retirement easy, a latent function was to reduce
the number of farmers and increase the scale of agriculture by
giving financial incentives both to any farmer who retired from
farming and to the third party who bought his land.
Acceleration of land transfer was promoted through the
provision that land must be transferred before the age of 60.
In order to qualify for benefits, "U-turn" heirs to farms (heirs
who migrate to the city for urban jobs and then return), must
return to the farm six months prior to the farm manager's land
transfer. In addition, enrollment prior to age 35 or
management of a farm with acreage over the prefectural
average entitled young farmers to a 30% reduction in premiums.
The premiums were relaxed for those entering the program
when it was started. By August 1973 over one million farmers
belonged to one of the programs. The three types of pension
programs are listed below:

1. The Farm Management Transfer Pension Fund (Keiei Ijō
   Nenkin): The pension benefits start from age 60 but are
   reduced to 1/10 the amount after age 65 when the Farmers
   Old Age Pension starts.
2. Farmers Old Age Pension (Nōgyōsha Rōreinenkin): The
   pension starts from age 65 to complement the national
   retirement system. See item 13 below for approximate
   yearly payments.

3. Program Eligibility:

Birth Date	Membership Period Necessary For Benefits
1916-1920	5 years
1921-1922	6 years
1923-1935	7 years to 19 years based on birthdate
1936-present	20 years

4. Discontinuation of Benefits:
   a. Procurement of over 10 ares of farmland after transfer
   b. Receiving, buying, or leasing of land after transfer
5. Conditions for Joining:
   a. Must belong to National Retirement System
   b. Must be an agricultural farm manager cultivating at least 50 ares of farmland or grassland (2 hectares in Hokkaidō)
   c. Must be 40 years of age
6. Optional Joining:
   a. Heirs (kōkeisha): If an heir joins before the age of 35 there is a 30% reduction in dues.
   b. Intensive Cultivation Managers: Must have over 30 ares but under 50 ares in own name (1-2 hectares in Hokkaidō) and perform more than 700 hours labor time per year including work in an intensive agricultural operation such as in a vinyl greenhouse. The following labor credits are given:

Crop	Hours	Crop	Hours
paddy rice	125	vegetables	
field rice	90	(equipped)	1800
wheat	55	(non-equipped)	345
sweet potatoes	90	ornamentals	345
beans	40	mulberry	120
orchards		feed and fodder	40
(equipped)	890		
(non-equipped)	320		

7. Agricultural Producer Corporation Members: The corporation must manage over 50 ares of farm land per member. The members must be permanent employees.
8. Automatic Withdrawal: A person can leave farming by selling all the land or by becoming employed in a company.

9. Optional Withdrawal From System: Land becomes less than 50 ares because of urbanization, through injury or disease, or by other reasons.

10. Dues Payment: For 1988 the dues were approximately 78,120 yen per year for the special rate (for heirs under 35 years old or managers of land over the prefectural land scale average or persons who perform at least 1500 hours in agricultural labor time). Otherwise, the regular dues were 109,560 yen per year.

11. Pension Payments: In 1983 the expected payout for 20 years participation was 892,300 yen for each year between ages 60 and 65 and the following after age 65:

National Retirement System	726,600 yen per year
Agricultural Retirement	
1. Farm Transfer Pension	89,400 yen per year
2. Old Age Pension	223,400 yen per year
Total	1,039,400 yen per year

Source: Nihon Nōgyō Nenkan Kankōkai 1984:342.

12. U-Turn Heirs: Heirs who have migrated from the farm and return to it must return to the farm 6 months prior to the land transfer and must have worked in agriculture a total of 3 years.

13. Financial Incentives: Buying land from pension participants entitles a buyer to low interest loans of 3% for 30 years or 3.5% for 25 years. As an incentive to quit farming a 620,000 yen lump sum payment (rinōkyūtsukin) is given. The conditions require that a person has farmed over 5 years and is over 20 years old.

1970 PROMOTION OF COMPREHENSIVE AGRICULTURAL POLICY

The policy goals were defined:

1. To create as many large-scale and highly efficient viable farms as possible and also to encourage the development of "production organizations." Part-time farmers were included in such organizations so that the scale of production could be enlarged.

2. To control rice production and to change the pattern of agriculture so that it would meet the changing demand for food, giving due consideration to regional comparative advantage in agricultural production.

3. To stabilize the prices of agricultural products so that in the long run they would fully reflect the market balances. It also provided for the rationalization of agricultural distribution and processing.

4. To ensure that the incomes and living standards of farmers were comparable to those in other sectors through the promotion of both agricultural incomes and non-agricultural incomes.

5. To encourage and aid the smooth flow of farmers out of agriculture.

6. To establish vigorous and pleasant rural communities by improving public facilities for production and life with emphasis on the conservation of the natural environment and meeting the increasing demand of the countryside for recreational purposes.

1969-1970 SMALL-SCALE RICE DIVERSION PROGRAM

This program entailed a 10,000 hectare target and incentives of 20,000 yen per 0.1 hectare for farmers to divert rice acreage to non-production of rice.

1970 LARGE-SCALE RICE DIVERSION PROGRAM

This program diverted 236,000 hectares of paddy and 118,000 hectares of non-paddy land to non-agricultural use. Incentive payments amounted to 81 yen per kilogram of rice or 35,000 yen per 0.1 hectare (on the average). The program became overly successful and 337,000 hectares were diverted. Only 22% of the total acreage was diverted to crops such as vegetables, feed crops, and pulses. Approximately 78% was fallowed.

1971 AGRICULTURAL MACHINE BANKS
(Nōgyō Kikai Ginkō 農業機械銀行)

Similar to the West German Maschinenringe, these cooperative tractor and other large-scale agricultural machinery institutions were designed to integrate large scale machinery into systematic use by the family farm. The machinery banks were administered through the agricultural cooperatives.

1971 THE ACT TO PROMOTE THE INTRODUCTION OF INDUSTRY INTO RURAL AREAS
(Nōson Chiiki Kōgyō Dōnyū Sokushin Hō 農村地域工業導入促進法)

This act gave tax breaks to companies which located in rural areas and provided loans to these companies from the Central Agricultural Bank.

1971-1975 RICE PRODUCTION CONTROL AND DIVERSION PROGRAM
(Inasaku Tenkan Taisaku 稲作転換対策)

Rice crop reduction targets (gentan) were given to each prefecture which in turn determined the allocation of crop fallowing or diversion for each farmer. Preferential payments were given to perennial crops which were less likely to return to rice production. Payments for fallowing were less. By 1973 52% of the total diverted area was fallowed and 48% diverted to other crops. At that time a farmer could get about 30,000 yen per 0.1 hectare for 3 years for simple fallowing while special diversion to perennial crops brought him 40,000 yen per 0.1 hectare for 5 years.

1972 REGIONAL AGRICULTURAL PRODUCTION INTEGRATION PROGRAM
(Nōgyō Danchi 農業団地)

A goal of joint use of facilities such as tractors and other equipment was established to promote systematic application of large-scale production techniques. The size of the integrated units was 40 hectares for rice, 50-100 hectares for wheat and barley, 300 head of cattle for dairy cows, 500 head of cattle for meat production, and 10 hectares for vegetable production. See "crop diversion programs" in 1978.

1973-1979 THE TOKYO ROUND OF THE GENERAL AGREEMENT ON TRADE AND TARIFFS (G.A.T.T.)
(Gatto Bōeki Kōshō Tōkyō Raundo ガット貿易交渉東京ラウンド)

The Tokyo Round, achieving broader concessions than the Kennedy Round, instituted a zero tariff binding on soybeans. This was the major item among the fourteen items to which tariff binding levels were applied. Henceforth, U.S. soybeans were imported duty free from the U.S. The value of Japanese concessions in the bilateral Japanese-U.S. package covering 150 agricultural items averaged a 35% price reduction. Import quota increases of 31% for beef, 82% for oranges, and 19% for orange and grapefruit juice were also a significant part of the tariff and quota adjustments which amounted to $211 million annually in 1976 dollars.

1975 AGRICULTURALLY PROMOTED LANDS LAW
(<u>Nōgyō Shinkō Hō</u> 農業振興法)

This law provided for the conversion of agricultural land to non-farm use. The agricultural committees oversee and mediate this conversion with the goal of paddy consolidation and rationalization. A definition of "<u>nōyochi</u>" and provisions for population density are given.

1976 COMPREHENSIVE PADDY FIELD UTILIZATION PROGRAM
(<u>Suiden Sōgō Riyō Taisaku</u> 水田総合利用対策)

This was a continuation of the Rice Production Control and Diversion Program.

1978 PROGRAM TO REORGANIZE THE UTILIZATION OF THE PADDY FIELDS
(<u>Suiden Riyō Saihen Taisaku</u> 水田利用再編対策)

This policy established a system of non-rice crop diversions. It replaces the <u>gentan</u> policy. Nevertheless, this policy is perceived by the farmers to be a continuation of the <u>gentan</u> policy and is therefore called <u>gentan</u> by them. This policy gives subsidies for growing specified crops on land formerly used for rice production and is commonly referred to as the rice diversion policy <u>tensaku</u>. While the <u>gentan</u> policy was unilaterally applied throughout the country, the <u>tensaku</u> policy has different quotas for each prefecture. The diverted acreage is as follows:

<u>Rice Land Diversion Acreage Goals</u>

<u>Program</u>	<u>Year</u>	<u>Acreage Diverted X 1000 Hectares</u>
Rice Production and Diversion	1971	541
Program (<u>gentan</u> policy)	1972	566
	1973	562
	1974	313
	1975	264
Comprehensive Paddy Field	1976	194
Utilization Program	1977	212

Program to Reorganize the
Utilization of the Paddy
Fields

Phase One	1978	438	
	1979	471	
	1980	584	
Phase Two	1981	666	
	1982	667	
	1983	630	
Phase Three	1984	557	
	1985	511	
	1986	600	
	1987	718	
	1988	770	

The 1987 crop diversion goals for each prefecture are listed as follows:

PREFECTURAL CROP DIVERSION ACREAGE (in hectares)

Prefecture	1987 Diverted Acreage (A)	1969 Planted Acreage (B)	A/B
Whole Nation	772,727	3,173,000	24%
Hokkaidō	126,633	266,200	48%
Aomori	25,231	90,400	28%
Iwate	23,834	98,500	24%
Miyagi	20,654	127,200	16%
Akita	29,714	123,300	24%
Yamagata	19,269	107,300	18%
Fukushima	25,282	115,100	22%
Ibaraki	27,390	113,800	24%
Tochigi	30,016	105,000	29%
Gumma	10,390	39,800	26%
Saitama	16,700	76,600	22%
Chiba	19,481	102,600	19%
Tōkyō	489	2,940	16%
Kanagawa	2,570	11,700	22%
Niigata	30,071	186,400	16%
Toyama	14,140	74,700	19%
Ishikawa	8,730	51,300	17%

Fukui	8,180	47,900	17%
Yamanashi	3,910	14,700	27%
Nagano	19,700	75,000	26%
Gifu	13,770	61,300	22%
Shizuoka	10,980	48,600	23%
Aichi	16,540	77,300	21%
Mie	12,920	67,500	19%
Shiga	10,820	62,200	17%
Kyōtō	6,460	34,900	19%
Ōsaka	4,820	21,400	23%
Hyōgo	22,530	89,000	25%
Nara	6,790	25,000	27%
Wakayama	4,600	21,200	22%
Tottori	8,010	30,200	27%
Shimane	8,150	48,100	17%
Okayama	16,520	77,700	21%
Hiroshima	13,620	64,200	21%
Yamaguchi	11,720	58,700	20%
Tokushima	7,470	27,800	27%
Kagawa	8,250	34,400	24%
Ehime	8,430	37,000	23%
Kōchi	10,940	35,900	23%
Fukuoka	22,630	90,300	25%
Saga	12,260	53,600	23%
Nagasaki	7,070	31,700	22%
Kumamoto	22,980	79,900	29%
Oita	11,610	52,800	22%
Miyazaki	14,301	48,300	30%
Kagoshima	16,160	63,900	25%

1987 Subsidies for Diverting to Non-Rice Crops: In addition to giving farmers a base amount for growing non-rice crops on plots formerly devoted to rice, supplements were awarded for increased agricultural productivity or district ventures. For example, items which qualified for agricultural productivity were increased scale of production, improved organization of production, formation of a danchi (see below), change-over to vegetable crops away from rice, or combining fodder crops with livestock. If the diversion was coordinated by Nōkyō or any other unit on a district basis, additional subsidies were available.

	Base Amount	Additional Supplements Productivity	District
Regular Crops soybeans, animal feed fodder, wheat buckwheat sugar beets	20,000 yen	20,000 yen	10,000 yen
Long Term Crops 5 yr. bearing orchards	25,000 yen	20,000 yen	10,000 yen
Special Crops vegetable tobacco konnyaku	7,000 yen	5,000 yen	5,000 yen
Fallowed Paddies	7,000 yen	-------	-------
Paddies in the Land Improvement Project	7,000 yen	-------	-------

Planned Additions to the above:
In the event that an area surpasses its crop diversion goal the following chart provides for additional subsidies. In order to qualify, at least one-half of the crop diversion plots must exceed one hectare or have at least two individuals per hamlet in a group called a "danchi."

Qualifications for Becoming Danchi Areas:
1. Contiguity: One-fourth the circumference of the diverted paddy must adjoin other such paddies; there must be irrigation ditches of at least 2 meters present; there must be 5 meter roads; and the diverted crop should be the same as other diverted crops in the area and be under 20% of the total crop diversion paddy area.
2. Scale:
   a. Monocropped Areas: 3 hectares or more of contiguous danchi plots.

b. Mixed-crop Areas: 1 hectare or more of contiguous danchi plots. The total of the respective plots must be at least 2/3 of the total crop diversion paddy area. (Note: For promoted mountain areas the amount is 0.7 hectares)
3. Crop Conformity: At least 90% of the danchi area must consist of 1 or 2 crops. In case of danchi that are 2 hectares or over, 3 crops are permitted.

The financing system available for crop diversion in 1983 is described below. Most funding is financed through Nōkyo (See Chapter 8).

1. The Agricultural Modernization Fund (Nōgyō Kindaika Shikin)
   This fund was established in 1961 to help modernize Japanese agriculture. All funds could be borrowed from Nōkyō or the prefectural trust bank federation.
   a. Construction: The construction fund is used for constructing barns, manure sheds, silos, facilities for storing agricultural products, mushroom cultivation facilities, and pollution prevention facilities. Farm managers can borrow up to 6 million yen at 6% interest for 12 years. Farm corporations may borrow up to 50 million yen at 7% interest for 12 years. Nōkyō or a non-agricultural enterprise which carries out a program which is for cooperative utilization qualifies to borrow up to 25 million yen at 7% for 15 years.
   b. Farm Machinery Fund: The farm machinery fund is for the purchase of generators, plows, sprayers, and harvesting implements. The conditions and who qualifies are the same as above except that the borrowing period is between 7 to 10 years.
   c. Orchard and Planting Fund: The orchard and planting fund is for the purchase of fruit trees, tea plants, and mulberry trees. The conditions are the same as above except that the borrowing period is 5 to 7 years.
   d. Small-scale Land Improvement Fund: This fund is for agricultural land or pasture improvement. Individual farmers can borrow up to 1.6 million yen at 5% interest for 10 years for each case. It should be remembered

that there is an additional 2% interest off if the above
are used to facilitate crop diversion.

2. Agricultural Improvement Fund (Nōgyō Kairyōshikin): This
fund was established in 1956, and in 1988   provided farmers
or farmer's groups low rate loans for 3-7 years. In 1987
the interest rate was 4.5% (Nihon Nōgyō Nenkan Kankōkai
1988:353).   The prefectural government sets the amount
allotted for each item. The fund, which is managed by
Nōkyō, can loan up to 80% of the prefectural figure and is
mainly used for energy saving techniques, instruction,
improvement of the production environment (for example,
livestock waste disposal), the promotion of managed crop
conversion (for example, facilities, machinery, and rent),
development of agricultural skills, and fostering of
production organizations (for example, joint utilization of
machinery facilities).

3. Funding is also available through the Agriculture, Forestry,
and Fisheries Capital Fund (Nōrin Kōko Shikin), which was
established in 1952. Most of the funds go towards financing
project related to the Land Improvement Project.

1980 THE REVISION OF THE AGRICULTURAL LAND LAW
(Nōchihō Kaisei Hō 農地法改正法)

The Program to Promote Agricultural Land Utilization was
exempted from the agricultural land law rules.   Land
Commission permission was not deemed necessary for
transactions although the program was carried out by the Land
Commission. Both the upper and lower limits on land holdings
were lifted and non-farmers could engage in agricultural
production corporations. Rental payments could be made in cash
(kinnōsei) or kind (butsunōsei). The reasoning for payment in
kind was to enable the aged and part-time farmers to directly
receive food as rent.

1980 ACT TO PROMOTE AGRICULTURAL LAND
UTILIZATION
(Noyōchi Riyō Zōshin Hō 農用地利用増進法)

This program has three sub-programs:

1. The Program to Promote the Establishment of Utilization
Rights (Riyōken Setteito Sokushin Jigyō): Its goal was to
facilitate the leasing of land.

2. The Program to Improve Agricultural Land Use (Nōyōchi Riyō Kaizen Jigyō): This program facilitated group crop conversion based on residential areas such as hamlets or villages and on land locations districts (ōaza). In either case two-thirds agreement of the households involved was necessary to qualify.

3. Project to Promote the Entrusting of Agricultural Production (Nosagyō Juitaku Sokushin Jigyō): This program supported agricultural machinery banks at agricultural cooperatives and agricultural production organizations.

A goal of affecting 5% of the total agricultural land over a period of 5 years was targeted in order to speed up land transfers, paddy consolidation and crop conversion. This program was called the Nōyōchi Kōdo Riyō Sokushin Jigyō. Below are listed the merits of the program as translated from a brochure obtained from a Land Commission office. Advantages to persons wanting to lease out or sell land are as follows:

1. Permission of the Land Commission is not necessary.
2. The lessor of land can have the land returned with the certainty of not having to pay compensation to the lessee.
3. The lessee receives a subsidy of 10,000 yen/10 ares if the duration of lease is between 3 and 6 years, or 20,000 yen/10 ares for periods of over 6 years. The subsidy for grass and pasture is 2000 yen and 4000 yen for the same periods.
4. A 5,000,000 yen deduction from the income tax land transfer profit is granted in the event of a land sale under the program.
5. The leased land will not be subject to the limit on land rented out and it is safe to lend out land outside the city, township, or village.
6. A 620,000 yen lump sum retirement subsidy is provided to program participants who stop farming and sell their land.
7. If the local program is tied to the prefectural agricultural development public corporation, the rent can be received in a lump sum in advance.

Advantages for the person who wishes to expand the scale of production by buying or leasing land are as follows:

1. The permission of the Land Commission is not necessary.

2. Land use conversions can include agricultural facilities such as barns or other facilities.
3. The lessee is guaranteed full use of the land during the contract period with the possibility that the contract will be renewed.
4. A borrowing limit of 15,000,000 yen is possible from the Land Purchase Fund. Normally the limit is 2,000,000 yen.
5. For land sales the title registration is reduced from 50/1000 to 9/1000.
6. Concerning the real estate purchase tax, one-third the purchase price is deductible. For lands designated as "agricultural use districts" or "agricultural promoted areas", one-fourth the purchase price is deductible.
7. Registration of the ownership rights is recorded by, but not controlled by, the Land Commission.
8. Special financing is available for the Program for Scale Enlargement and the Introduction of Cooperative Utilization of Facilities Projects.

1980 THE BASIC DIRECTION OF AGRICULTURAL POLICY FOR THE 1980'S: A Report by the Agricultural Policy Deliberation Council October 31.
(Hachiju Nendai no Nosei no Kihon Hoko 80年代の農政の基本方向)
This major policy statement charted self-sufficiency goals in Japanese agriculture, a Japanese-style diet, and fostering of a more efficient, larger-scale agriculture. The major points addressed are translated as follows:

1. In 1975 the rural population comprised 40% of the total Japanese population. In 1990 it will make up only 30% of a projected population of 130 million people.
2. Agriculture and the production of food, besides forming an important part of the country's economy and society, constitute a link in the chain of security that guarantees the life of the nation's people.
3. Henceforth, agricultural policy, together with meeting the above projections and evoking the will of the farmers, should make a ceaseless effort to stabilize the food supply for the nation's 100 million people. For this purpose it is necessary to place special emphasis on the following points:

a. To enhance the nation's food self-sufficiency. This should include a guarantee that agricultural land be preserved as central towards this end.
b. To further the reconstitution of agricultural productivity to meet consumer demand.
c. To plan for increased agricultural productivity.
d. To intentionally equip rural villages as local societies with plentiful greenery. (For example, maintain strict zoning ordinances that limit industrial development).
e. To map out the establishment of a food supply system for the production of food items which conform to the trend of concerns and food habit preferences of the nation's people.

In order to accomplish the above, the following concepts were promoted:

1. Japanese-style food habits (Nihongata Shokuseikatsu)
2. Core farmers (chūkaku nōka) who will take a central role in the agricultural production. Note: The idea of "core farms" dates to the early 1960s and the Agricultural Basic Law. They are defined as farms which have a male heir under age 60 who works over 150 days a year on his/her own farm.
3. Enlarging the scale of core farms by adding land from the Type II part-time farmers and retired farmers who sell, lease, or entrust their farms.
4. Maintenance of the rice supply while balancing the rice surplus through crop diversion.
5. Increasing the agricultural productivity while developing part-time farming, mixed farming, and industrial residential areas. The Third Comprehensive National Development Plan, Sanzensō, fostered a new planning concept utilizing residential areas defined by population density and industrial development. These units, listed below from small to large do not necessarily coincide with existing administrative units.

a. Kyojūku, which includes the hamlet unit shūraku.
b. The former city, township or village unit teijuku.
c. Several cities, townships, or village units teijuken.

1980 THE LONG TERM FORECAST OF THE DEMAND AND PRODUCTION OF AGRICULTURAL PRODUCTS
(Nōsanbutsu no Juyō to Seisan no Chōki Mitōshi 農作物の需要
と生産の長期見通し)
See Tables A.1, A.2, A.3, and A.4

Table A.1
Food Self-Sufficiency Percentages

	1978	1990
Composit Food Self-Sufficiency	73%	73%
Grain Staples Self-Sufficiency (A) (excluding feed grain)	68%	68%
Feed Grain Self-Sufficiency (B)	29%	35%
Grain Self-Sufficiency (A+B)	34%	30%

1984 U.S.-JAPAN AGRICULTURAL TRADE AGREEMENT
(Nichibei Gōi Kakunin Bunshō 日米合意確認文章)
Japan agreed to increase quotas of high-grade beef and citrus
over the next four years. The agreement expired in March
1988.

Year	Beef	Orange	Concentrated Fruit Juices Orange	Grapefruit
1983	141,000	82,000	6,500	6,000
1984	150,000	93,000	7,000	--
1985	159,000	104,000	12,500	--
1986	168,000	115,000	8,000	--
1987	214,000	126,000	8,500	--

(in metric tons)

1986 THE BASIC DIRECTION OF AGRICULTURAL POLICY AIMING AT THE 21ST CENTURY: A Report by the Agricultural Policy Deliberation Council November 28.
(Nijuisseki e Mukete no Nōsei no Kihon Hōkō 21世紀へ向
けての農政の基本方向)
This was a revision of the 1980 planning guide. The major
changes are to:

1. Establish high productivity paddy-field farming based on
   lower labor inputs through increased mechanization of a
   larger land scale. The goal for large-scale mechanized

Table A.2
The Rice Supply and Its Forecast

	1968	1977	1978	1979	1980	1990
Rice Consumed Per Person	100	83	82	80	79	63–66
Rice Demand (X 10,000 tons)	1,225	1,148	1,136	1,112	1,121	970–1,020
Rice Supply (X 10,000 tons)	1,445	1,310	1,310	1,196	975	1,000

Note: Based on the assumption that in 1990 there will be 2.72 million hectares of paddy rice and that 0.76 million hectares will be diverted to other crops.

Table A.3
Per Capital Consumption and Production of Agricultural Products 1978-1990

Crop Type	1978 Per Capita (kilograms)	1978 Total (10,000 tons)	1990 Per Capita (Kilograms)	1990 Total (10,000 tons)
Rice	81.6	1,136	63-66	970-1,020
Wheat	31.7	586	32	641
Oats and Barley	0.7	238	0.4	348
Sweet Potatoes	4.2	137	4.3	132
Potatoes	13.7	356	15	386
Soybeans	5.3	419	5.4	520-543
Peanuts	0.8	16	0.8	18
Other Beans	2.3	29	2.2	32
Vegetables	114.9	1,686	114	1,826
Tangerines (Mikan)	16.2	321	16	354
Apples	5.4	84	6.4	110
Other Fruit	19.5	392	21	472
Milk and Dairy Products	59.3	701	71-75	927-972
Beef	3.3	56	4-5	85-92
Pork	8.7	147	11	196-210
Chicken	7.1	109	8-9	147-155
Other Meat	1.7	36	1-2	44-46
Eggs	14.9	204	15	225
Sugar	24.8	292	25	321
Oils and Fats	12.8	191	16-17	252-266
Tea	0.9	11	0.92	12

Source: Noseishingikai 1980.

Table A.4
Food Consumption and Production Forecast and Projected Self–Sufficiency Rates

Crop	Domestic Consumption		Domestic Production		Self–Sufficiency	
	1978	1990	1978	1990	1978	1990
Rice	1,136	970–1,020	1,259	1,000	111	100
Wheat	586	641	37	122	6	19
Oats & Barley	238	348	33	58	14	17
Vegetables	1,686	1,826	1,641	1,826	97	99
Fruit	790	921	616	768	78	83
Milk & Milk Products	701	927–972	626	842	89	89
Beef	56	85–92	41	63	73	71
Pork	147	196–210	132	194	90	95
Chicken	109	147–115	102	146	94	96
Eggs	20	225	198	222	97	99
Sugar	292	321	67	102	23	32

Source: Nōseishingikai 1980.

farms was 30 to 45 hectares and the goal for medium-scale mechanized farms was 12 to 24 hectares.

2. Take emphasis away from subsidies for rice crop diversion (see "1978 Program to Reorganize the Utilization of the Paddy Fields") and instead institute incentives to be used for structural improvement of rice productivity.

3. Improve the food control system by selling off excess rice.

4. Aid the establishment of agriculture as a highly efficient self-dependent industry.

5. Develop a commodity pricing system to provide lower stable prices.

6. Cooperate with GATT rules to establish a new worldwide order in agricultural trade.

7. Develop new technologies related to electronics and biotechnology.

1988 U.S.-JAPAN AGRICULTURAL TRADE AGREEMENT
(Nichibei Gōi Kakunin Bunshō 日米合意確認文章)

In return for the U.S. withdrawing its complaint to GATT concerning the legality of Japan's beef and citrus quotas, beef import quotas were to be gradually removed by April 1, 1991 by raising the 1987 level by 60,000 metric tons per year. Japan was permitted to levy tariffs on imported beef in the amounts of 70% in FY 1991, 60% in FY 1992, and 50% from April 1993. The issue of the level of future beef tariffs was negotiable. Orange imports were likewise liberalized by 1991 by increasing the 1987 level by 22,000 metric tons per year. After 1991 the tariffs were to remain at 20% in the off-season and 40% during the remainder of the year. Tariffs were lowered on other commodites including grapefruit, lemons, frozen peaches and pears, pistachios, macadamias, pecans, walnuts, bulk pet food, pet food for retail sale, beef jerky, sausage, and pork and beans. Orange juice imports were to be freed by April 1, 1992, the year after oranges are liberalized.

Appendix B:
Measurements

Square Measure
Japanese	American	Metric	
10 tan = 1 chō	= 2.45 acres	= 99.2	ares
		= 0.992	hectares

Capacity
10 shō = 1 to	= 4.8 gallons	= 18	liters
10 to = 1 koku	= 44.8 gallons	= 180	liters

Rice Weights
1000 momme = 1 kan	= 8.72 pounds	= 3.75	kilograms
1 koku brown rice	= 330 pounds	= 150	kilograms
	100 pounds	= .0454	metric tons
	48.99 bushels	= 1	metric ton
	1 cwt rough	= .032	metric tons milled

Currency
125 yen   = $1.00

Bibliography

WORKS IN ENGLISH

Australian Bureau of Agricultural and Resource Economics (A.B.A.R.E.)
  1988 Japanese Agricultural Policies: A Time of Change. Policy Monograph No.3. Canberra, Australia: ABARE.
Bachnik, Jane M.
  1981 Recruitment Strategies for Household Succession: Rethinking Japanese Household Organizations. Paper presented at 1981 Annual Meeting of the American Anthropological Association. Ann Arbor: The University of Michigan.
Bix, Herbert
  1987 Class Conflict in Rural Japan: On Historical Methodology. Bulletin of Concerned Asian Scholars 19(3):29-42.
Borgstrom, Georg
  1967 The Hungry Planet. New York: Macmillan.
Brandt, Vincent S.R.
  1971 A Korean Village. Harvard East Asian Series 65. Cambridge: Harvard University Press.
Brow, James
  1981 Class Formation and Ideological Practice: A Case from Sri Lanka. Journal of Asian Studies 40(4):703-718.
Brown, Keith
  1968 The Context of Dozoku Relationships in Japan. Ethnology 7:113-138.
Carlson, Eugene
  1987 Japanese Companies Increase Presence Near Mexico Border.

The Wall Street Journal, December 22.

Chayanov (Tschajanov), A.V.
1966 (original 1925) The Theory of Peasant Economy. (Edited by Daniel Thorner, Basile Kerblay, and R.E.F. Smith). Homewood, Illinois: Irwin.

Chen, Chung-min
1977 Upper Camp: A Study of a Chinese Mixed Cropping Village in Taiwan. Taipei: Republic of China.

Chira, Susan
1982 Cautious Revolutionaries: Occupation Planners and Japan's Postwar Land Reform. Agricultural Policy Research Center Publication. Tokyo: Nihonkei.
1988 Leaner Japanese Manufacturers. New York Times, February 18.

Cole, Allan B.; George O. Totten, and Cecil H. Uyehara
1966 Socialist Parties in Postwar Japan. New Haven, Conn.: Yale University Press.

Cornell, John B.
1963 Local Group Stability in the Japanese Community. Human Organization 22:113–125.

Dore, Ronald P.
1959 Land Reform in Japan. New York: Oxford University Press.

Ellwood, Robert S.
1973 The Feast of Kingship: Accession Ceremonies in Ancient Japan. Monumenta Nipponica, Sophia University, Tokyo.

Firth, Raymond
1951 Elements of Social Organization. London: Watts.

Francks, Penelope
1984 Technology and Agricultural Development in Pre-War Japan. New Haven, Conn.: Yale University Press.

Freedman, Maurice
1965 Lineage Organization in Southeastern China. Monographs on Social Anthropology: No.18. London: Athlone Press.

Fuji Bank Ltd., Research Division
1983–1988 Fuji Bank Bulletin. Tokyo.

Fukutake, Tadashi
1980 Rural Society in Japan. Tokyo: University of Tokyo Press.

Grad, Andrew
1952 Land and Peasant in Japan: An Introductory Survey. ^New York: International Institute of Pacific Relations.

Hayami, Yūjiro
1975 A Century of Agricultural Growth in Japan. Tokyo:

University of Tokyo Press.
1988 Japanese Agriculture Under Siege. Houndsmills, Basingstoke, and Hampshire, England: Macmillan Press.
Hewes, Lawrence I. Jr.
1950 Japanese Land Reform Program. Report No.127, SCAP, Natural Resources Section.
Holtom, D.C.
1972 (original 1928) The Japanese Enthronement Ceremonies. Monumenta Nipponica: Sophia University, Tokyo.
Huang, Shu-min
1982 Agricultural Degradation: Changing Community Systems in Rural Taiwan. Washington, D.C.: University Press of America.
Ishino, Iwao and John Donoghue
1964 Small Versus Large Scale Agriculture. Human Organization 23(2):119-122.
Izumi, S.; C. Ogyu; K. Sugiyama; H. Tomoeda; and N. Nagashima
1984 Regional Types in Japanese Culture. In Regional Differences in Japanese Culture. Nagashima and Tomoeda, eds. Senri Ethonological Studies No.14:187-198.
Jones, Randall
1987 The Uruguay Round of Multilateral Trade Negotiations. Japan Economic Journal No.22A, June 12.
Keizai Kōho Center
1987 Rice Deregulation: A Multi-stage Approach. No. 41, March. Tokyo.
Kelly, William
1982(a) Irrigation Management in Japan: A Critical Review of Japanese Social Science Research. Cornell East Asia Papers Number 30. Cornell China-Japan Program and Rural Development Committee.
1982(b) Water Control in Tokugawa Japan: Irrigation Organization in a Japanese River Basin, 1600-1870. Ithaca: Cornell University East Asia Papers.
Kenmochi, Kazumi
1987 The Hollowing. Ampo: Japan-Asia Quarterly Review Vol.19, No.1:30-33.
King, F.H.
1911 Farmers of Forty Centuries: Permanent Agriculture in China, Korea, and Japan. Mrs. F.H. King: Madison Wisconsin.
Kitazawa, Yōko
1987 Setting Up Shop Shutting Up Shop. Ampo: Japan-Asia

Quarterly Review Vol.19, No.1:10-29.

Kunihiro, Narumi
1984 Family Ties in an Urban Era. Japan Echo 11(2): 80-84.

Lee, kwang-kyu
1976 A Comparative Study of the Rule of Descent in East Asia: China, Korea, and Japan. Korea Journal 16(11): 12-22.

MacKnight, Susan
1987 Japan, Agriculture, and the MTN. Japan Economic Report No.44A, November 20.
1989 The U.S. Trade Deficit with Japan: "Something Must Be Done. Japan Economic Report No.17A, April 28.

Mao Yu-kang
1982 Land Reform and Agricultural Development in Taiwan. In Chi-ming Hou and Tzong-shian Yu, ed., Agricultural Development in China, Japan, and Korea. Taipei, Taiwan: Academia Sinica.

Midoro, Tatsuo
1982 Agricultural Cooperatives and Special Purpose Agricultural Cooperatives in Japan. Gakujutsu Kenkyu Hokoku No. 14. Nayoro Joshi Tanki Daigaku (Nayoro Women's Junior College).

Miner, William M.
1987 Foreign Aspects of Decoupling: The Use of Measurements in Relation to Decoupling and Trade Negotiations, Institute for Research on Public Policy, Ottawa, January (mineo).

Miner, William M. and Dale E. Hathaway, Eds.
1988 World Agricultural Trade: Building a Consensus. Halifax, Nova Scotia: The Institute for Research on Public Policy.

Moore, Richard H.
1985 Land Tenure and Social Organization in a Rice Growing Community in Tohoku Japan. Dissertation at the University of Texas at Austin.
1986 Japanese Rural Subcontracting and Its Relationship to Social Organization. Papers in East Asian Studies No.1, East Asian Studies, The Ohio State University.

Nagashima, Nobuhiro and Hiroyasu Tomoeda, ed.
1984 Regional Differences in a Japanese Rural Culture: Results of a Questionnaire. Senri Ethnological Studies No.14. National Museum of Ethnology.

Nakamura, Masanori
1988 The Japanese Landlord System and Tenancy Disputes: A Reply to Richard Smethurst's Criticism. Bulletin of Concerned Asian Scholars 20(1):36-50.

Nakane, Chie
1967 Kinship and Economic Organization in Rural Japan. London School of Economics Monograph on Social Anthropology 32. London: Althone.
Nash, Nathaniel C.
1988 Japan's Banks: Top 10 in Deposits. The New York Times July 20.
Norinsuisansho (Ministry of Agriculture, Forestries, and Fisheries)
1979 Stable Supply of Foods and the Role of Agriculture. Tokyo.
Ogura, Takekazu
1982 Can Japanese Agriculture Survive?   Revised Edition. Agricultural Policy Research Center. Tokyo: Kyodo Printing.
Okawa, Kazushi
1982 Agricultural Development in Sectoral Interdependence: Views Derived from Japan's Experience. In Agricultural Development in China, Japan, and Korea. C.M. Hou and T.S. Yu, eds.   Taipei: The Institute of Economics, Academia Sinica.
Olson, Gary L.
1974 U.S. Foreign Policy and the Third World Peasant: Land Reform in Asia and Latin America. New York: Praeger.
Olstrom, Douglas
1988 Japanese Banks in the United States.   JEI Report January 22.
Otohiko, Hasumi
1985 Rural Society in Postwar Japan.   The Japan Foundation Newsletter (Feb.) Vol.12, No.5 and 6.
Polanyi, Karl
1957 The Economy as Instituted Process. In Trade and Market in the Early Empires. K. Polanyi et al., eds., pp. 243-270. Glencoe, N.Y.: The Free Press.
Republic of China, Bureau of Statistics.
1955-1988 Taiwan Agricultural Yearbook. Taipei.
1987 Statistical Bureau of the Republic of China 1987. Taipei.
Republic of Korea, Bureau of Statistics.
1955-1988 Korea Statistical Yearbook. Seoul.
Rostow, Walter
1951 The Stages of Economic Growth. Cambridge: Cambridge University Press.
Sahlins, Marshall
1972 Stone-Age Economics.   Chicago: Aldine-Atheron.

320

Scott, James
   1976 The Moral Economy of the Peasant. New Haven: Yale
         University Press.
Smethurst, Richard
   1986 Agricultural Development and Tenancy Disputes in Japan.
         Princeton: Princeton University Press.
   1983 Japanese Society. New York: Cambridge University Press.
Smith, Thomas C.
   1959 Agrarian Origins of Modern Japan. Stanford: Stanford
         University Press.
Sorensen, Clark
   1988 Over the Mountains are Mountains: Korean Peasant
         Households and their Adaptations to Rapid
         Industrialization. Seattle: University of Washington Press.
Suenari, Michio
   1972 (a)Yearly Rituals within the Household: A Case Study from
         a Hamlet in Northeastern Japan. East Asian Cultural
         Studies 11.
   1972 (b)First Child Inheritance in Japan. Ethnology 11:122-126.
Tanaka, Kakuei
   1973 Building a New Japan: A Plan for Remodeling the Japanese
         Archipelago. Tokyo: Simul Press.
Tsuchiya, Keizo
   1976 Productivity and Technological Progress in Japanese
         Agriculture. Tokyo: University of Tokyo Press.
Turner, Victor
   1967 The Forest of Symbols: Aspects of Ndembu Ritual. Ithaca,
         New York: Cornell University Press.
United States Department of Agriculture
   1982 Agricultural Statistical Yearbook. Washington, D.C.
         Government Printing Office.
United States Department of Agriculture, Economic Research
   Service
   1988 Rice: Situation and Outlook Report. RS-53 October 1988.
         Washington, D.C.: U.S. Government Printing Office.
Walinsky, Louis J. (Ed.)
   1977 Selected Papers of Wolf Ladejinsky: Agrarian Reform as
         Unfinished Business. Oxford: Oxford University Press.
Waswo, Ann
   1977 Japanese Landlords: The Decline of a Rural Elite.
         Berkeley: University of California Press.
Wolf, Margery
   1972 Women and the Family in Rural Taiwan. Stanford:

Stanford University Press.

Yamazaki, Hiromi
1987 Japan Imports Brides From the Philippines: Can Isolated Farmers Buy Consolidation? Ampo: Japan-Asia Quarterly Review, Vol.19, No.4: 31.

Yoshioka, Yutaka
1982 The Personal View of A Japanese Negotiator. In U.S.-Japanese Agricultural Trade Relations. Emery Castle and Kenzo Hemmi, eds. Washington, D.C.: Resources for the Future, Inc.

WORKS IN JAPANESE

Aoki, Osamu
1986 Kenkyu Nōto--Meiji--Taishō--Shōwa Senzen ni okeru Kengyō Nōka no Dōko: Tōkeiteki Kansatsu ni Yoru Tokuchō (Research Notes on the Trends of Part-time Farming in the Prewar Meiji, Taishō, and Shōwa Periods). Nōgyō Keizai Kenkyūshi (Tohoku University Institute for Agricultural Research) 7: 37-43.

Aoki, Osamu, Shibuya Chōsei, and Arai Satoshi
1986 Meiji, Taishō, Shōwa Senzenki ni okeru Kengyō Nōka no Dōko (Changes in Part-time Farming Households during the Meiji, Taishō, and Prewar Shōwa Periods). Nōgyō Keizai Kenkyūshi 7:37-43.

1987 Miyagi-ken and Yamagata-ken in Keizai Kikakuchō Chōsakyoku (Economic Planning Office, Survey Bureau), In Endaka o Norikae: Aratana Hatten wo Mezasu Chiiki Keizai (Overcoming the Yen Appreciation: Regional Economics Aimed at New Developments). Tokyo: Ōkurashō Insatsu.

Ariga, Kizaemon
1943 Nihon Kazoku Seido to Kōsaku Seido (Japanese Family and Tenancy Systems). Tokyo.

Chūko Kigyōchō (Ed.)
1987 Chūko Kigyō Hakushō (White Paper on Small and Medium-Sized Industries). Tokyo: Ōkurashō.

Furuzawa Hiroshi
1988 Kome Aguribijinesu no Kikenna Senryaku (The Dangerous Strategy of American Agribusiness). Asahi Jānaru Vol.30, No.47:91:94.

322

Honda, Yukio
1982 Suiden wa Chikyū o Sukuu (Rice Paddies Will Save the Earth). Tokyo: Ie No Hikari.

Hoshi, Makoto
1975 Sengo Nihon Shihonshugi to Nōgyō Kiki no Kōzō (The Structure of Postwar Japanese Capitalism and Crises in Agriculture). Tokyo: Ocha No Mizu Shobō.

Isobe, Toshihiko et al., eds.
1982 Nihon Nōgyō no Kōzō Bunseki: 1980 Sekai Nōrin Sensasu (An Analysis of the Structure of Japanese Agriculture). Tokyo: Nōrin Tōkei Kyōkai.

Isobe, Toshihiko
1979 Nihon no Nōka. Tokyo: Nōrin Tōkei Kyōkai.
1985 Nihon Nogyo no Tochi Mondai. Tokyo: Tokyo Daigaku Shuppankai.
1988 Kazokusei Nōgyō no Bunseki Kadai (Topic Analysis of Family Farming--A Book Review of Shiina's Famiri Famu no Hikakushiteki Kenkyū). Tochi Seido Shigaku 30(3):58-64.

Kahoku Shimpōsha Henshūkyoku ed.
1981 Kome o Dōsura: Shokkan ga Kiota. Tokyo: Sanichi Shobō.
1987 Yuragu Kome: Ikiru Michi (Quivering Rice: A Path for Survival). Tokyo: Kahoku Shimpōsha.

Kajii, Kō
1982 Nihon Nōgyō Saihen no Senryaku (The Strategy of Japanese Agricultural Reorganization). Kashiwa Shobo.

Kajii, Tsutomu
1982 Nihon Nōgyō Saihen no Senryaku (A Strategy for Restructuring Japanese Agriculture). Tokyo: Kashiwa Shobō.

Kano, Yoshikazu
1987 Kome o Dō Suru: Nōsei Kaikaku no Kokoro (What Should be Done About Rice: The Spirit of Agricultural Policy Reform). Tokyo: NIhon Keizai Shinbunsha.

Katō, Tatsuo, ed.
1981 Seisan Soshiki to Kibo no Keizai (The Economics of Product Organization and Scale). No. 134 Nihon no Nōgyō: Asu e no Ayumi. Tokyo: Funi Shuppansha.

Kawaguchi, Akira
1983 Mura no Ryōiki to Nōgyō (Farming and the Territory of Villages). Tokyo: Ie No Hikari.

Kawai, Kazushige, ed.
1971 Daikibo Inasaku Keiei (Tadano Nōjō) no Kaitai to Nōchi

Hoyū Gorika Sokushin Jigyō (The Dissolution of the Large Scale Rice Management and Project to Rationalize the Agricultural Land Holdings of the Tadano Farm). Tochi to Nōgyō (Land and Agriculture) 1:56-63.

1983(a) Kengyō Nōka no Seikaku to Nōgyō Hatten no Hōkō (The Direction of Agricultural Development and Characteristics of Part-time Farming Households). Nōson Bunka Undō No.89: Nōsangyoson Bunkakyōkai.

1983(b) Kenkyū Notō: Kengyō Nōka Sekishutsu no Seisanryoku Kōzō (Research Notes on The Structure of the Productive Capacity of Part-time Farming Households). Nōgyō Keizai Kenkyūshi (Journal of Agricultural Economic Research) 4:22-33. Agricultural Economics Research Office of the Tohoku University Institute for Agricultural Research.

1987 Kome ga Abunai: Mugi to Onaji Michi wo Ayumōtoshiteiru (The Rice Situation is Dangerous: About to Tread the Same Path as Wheat). Dō Suru: Shoku to Nō (What Should be Done About Rice and Food?). Sendai: Miyagi Chiiki Jiji Kenkyūkai.

Keizai Kikakuchō Sōgō Keikakukyoku (Economic Planning Office, Comprehensive Planning Bureau, ed.)

1987 Nijuisseki e no Kihon Senryaku (Basic Strategies for the 21st Century). Tokyo: Tōyō Keizai Shinpōsha.

Kinoshita, Akira

1956 Nakadanuma (Nakada Swamp). Tōhoku Daigaku Keizai.

Kirishitan Shūmon Aratamechō 1880 and 1854 Kirishitan Shūmon Aratemechō (Cadastral Surveys on the Christians). Sakuraba Mura, Tome-gun, Miyagi-ken.

Kitakamigawa Engan Nakadachiku Tochi Kairyōku Sōmuka, ed.

1982 Kitakamigawa Engan Nakada (The Kitakami River Shore Nakada). Sanuma: Kawauchi.

Kitamura, Toshio

1950 and 1973 Nihon Kangai Suiri Kankō no Shiteki Kenkyū (The Historical Study of Customary Rules on the Utilization of Irrigation Water in Japan). 2 Vols. Tokyo: Iwanami.

Kiuchi, Nobutane

1988 Nihon no Shudō de Gatto no Kaikaku wo (A Restructuring of Gatt under Japan's Leadership). Chūō Kōron Vol. 103, No.12:64:72.

Kokudochō (Land Office), ed.

1987 Daiyonji Zenkoku Sōgō Kaihatsu Keikaku (The Fourth National Comprehensive Development Plan). Tokyo: Ōkurashō Insatsu.

324

Kome Mondai Kenkyūkai
  1981 Kome: Shokukanseiri Seido no Sugata (Rice: The Shape of
    the Food Control Agency System). Tokyo: Sōzō Shobō.
Kondō, Yasuo
  1981 Nōyōchi Kakuho to Kokudo Seisaku (National Land Policy
    and the Maintenance of Land for Agricultural Use). Tokyo:
    Ochanomizu Shobō.
Miyagi-ken Hasama Nōgyō Kairyō Fukyūsho
  1980-1987 Tome-gun Nōgyō Saizensen (Tome County Agriculture
    Front Line). Sanuma.
Miyagi-ken Kishō Saigai Hasama Chihō Taisaku Kaigi
  1981 Nōsakubutsu Ijō Kishō Saigai no Kiroku (The Record of
    Calamities Caused by Unusual Weather Conditions).
    Tsukidate: Nambuya Insatsu.
Miyagi-ken Nōgyō Kaigai (Miyagi Prefecture Land Commissions)
  1982 Inasaku Shidō Shishin To Gijutsu Taisaku Shōwa 57 (1982
    Policies For Technology and the Promotion of Rice
    Growing). Sendai.
Miyagi-ken Nōgyō Kaigi (Miyagi Prefecture Land Commissions)
  1981 Hikasegi Rōdōsha no Shūro Jittai ni kan Suru Chōsa Kekka
    (Survey Results Regarding the Actual Situation of the
    Employed Day and Seasonal Laborers). Miyagi Prefecture.
  1982-1983(a) Nōchi Jōsei (Agricultural Land Affairs). Sendai.
  1982-1983(b) Shōwa 57 Nendo Jōjōteki ni Tasangyo ni Jūji
    Shiteiru Kōkeisha no Ikō ni Kan Suru Chōsa Kekka
    (Survey Results Regarding the Opinions of (Farm)
    Successors Permanently Employed in Other Industries).
    Sendai.
  1985 Nōgyō Rōchin to ni Kansuru Chōsa Kekka (Survey Results
    Regarding Agricultural Wage Rates). Sendai.
  1986 Tahata Baibai Tō ni Kansuru Chōsa Kekka (Survey Results
    Regarding Land Prices of Paddies and Fields Shōwa 61).
    Sendai.
  1987 Wakai Nōka Kōkeisha no Shūgyō ni Kansuru Chōsa Kekka
    (Survey Results Regarding the Employment of Young Farm
    Successors Shōwa 62). Sendai.
Miyagi-ken Nōgyō Keiei Kenkyūkai, Toshizō Hayashi, ed.
  1966 Nōka to Nōkyō Ni Kan Suru Rekishi (The History of
    Farming Households and the Agricultural Cooperative).
    Sendai: Miyagiken Nōkyō.
Miyagi-ken Nōseibu (Agricultural Policy Division of Miyagi
    Prefecture)
  1984 Miyagi no Nōgyō (Agriculture in Miyagi). Sendai: Miyagi-
    ken Nōseibu.

Miyagi-ken Shōkōkai Rengōkai (Miyagi Prefecture Federation of Chambers of Commerce)
1980 Shōkibo Jigyō Taisaku Tokubetsu Suishin Jigyō Chōsa Kenkyū Hōkokusho (Survey Research Report on the Project to Promote the Special Countermeasures for Small-scale Projects). Sendai.
1981 Shōwa 55 Nendo shōkibo Jigyō Taisaku Tokubetsu Suishin Jigyō Chōsa Kenkyū Hōkokusho (1980 Research Report on the Survey of the Special Project to Promote Countermeasures for Small-scale Enterprises).
Miyagi-ken Tōhoku Nōseikyoku Tōkei Seihōbu, ed.
1981 Miyagi Nōrinsuisan Tōkei Nenpō S55-S56 (1980 Agricultural Yearbook for Miyagi Prefecture). Sendai: Miyagi Nōrin Tōkei Kyōkai.
Miyagi-ken, Tome-gun, Nakada-chō
1980 Nōson Sōgō Seibi Keikakusho (Master Plan for Farming Villages). Nakada Township.
1983 Shokuryō Jukyū Sōgō Taisaku no Aramashi (Outline of the Comprehensive Program on the Food Supply and Demand). Sendai: Miyagi-ken Nōseibu Shokuryō Jukyū Sōgō Taisakushitsu.
Miyagi-ken, Tome-gun, Nakada-chō Nōgyō Iinkai
1980 Shūrakuteki Riyō Keikaku Kuiki Settei Shiryō (Report on the Condition and Planning of Land Utilization by Nakada Hamlets). Nakada Township.
Miyagi-ken, Tome-gun Nakada-chō Sangyōka
1983 Ikōchōsa ni Miru "Nōgyō no Genjō to Hōkō" ("The Situation and Direction of Agriculture: A Survey"). Nakada Township.
Nagata Keijurō
1971 Nihon Nōgyō No Suiri Kōzō (The Structure of Irrigation In Japanese Agriculture). Tokyo: Iwanami Shoten.
1979 "Kōzō Hendōki No Nōgyō Yōsui Juyō Mondai" (Problems of Agricultural Water Demand in a Period of Structural Transition). In Mizu To Nihon Nōgyō (Water and Japanese Agriculture). Ogata Hiroyuki, ed. Tokyo: Tokyo Daigaku Shuppansha.
Nakada-chō
1892 Kiriezu. Nakada.
1955-1988 Kōhō Nakada (formerly Nakada Chōsei Dayori). (Township newspaper) Nakada-chō Nōgyōsha Seinen Kaigi.
1982 Nakada-chō Kokudo Riyō Keikaku. Nakada-chō.
1983 Tochi Daichō (Unofficial copy).

Nakada-chō Nōgyōsha Seinen Kaigi
1976-1978 Nakada-chō Nōgyosha Seinen Kaigi Teigenshū
(Statements by the Nakada Young Persons Committee on
Agriculture: Proceedings on the Yearly Studies and
Experiences Concerning the Project to Reorganize the
Nakada District). Nakadachiku Nōgyō Kibanseibijigyō Ni
Tsuite: Nenkan Gakushū Kiroku.

Nakada-chō Sangyōka
1983 Ikō Chōsa ni Miru: Nōgyō no Gendai to Hōkō. Nakada-
chō.

Nakada-chō Shōkōkai
1983 Chiiki Kouri Shōgyō Kindaika Taisaku Chōsa Hōkokusho
(Survey Report on the Plan to Modernize Regional
Retailing). Nakada Township.
1986 Nakada-chō Shōiki Shōgyō Shindan Hōkokusho (Diagnostic
Report on Nakada Township's Merchant Business). Miyagi
Prefecture.

Nakada-chō Shōkōkai Mura Okoshi Jigyō Jikkō Iinkai
1986 Muraokoshi Jigyō Hōkokusho (Report on the Village
Promotion Project). Nakada-chō Shōkōkai.

Nakada-chō Shōkōkai, Nakada-chō Shōkōkai Fujinbu
1985 Nakada-chō nai Kōba Gaikyō Chōsa Hōkoku (Research
Report on the Situation of Factories In Nakada Township).
March.

Nakada-chō Tochi Kairyōku
1983 Aza Zumen (District Maps). Nakada-chō.

Nakada Chōshi Henshū Iinkai, ed.
1970-1977 Nakada Chōshi Shiryōshū (Collected Resource
Materials on the History of Nakada Township). 8 volumes.
Nakada Township.
1977 Nakada Chōshi (The History of Nakada Township). Sendai:
Ono Insatsusho.

Nihon Keizai Shinbun
1983 Tochi Kairyō Keikaku (The Land Improvement Plan). Nihon
Keizai Shimbun. April 14.

Nihon Nōgyō Nenkan Kankōkai
1983-88 Nihon Nōgyō Nenkan (Japan Agricultural Yearbook).
Tokyo: Ie No Hikari.

Nōchi Kaikaku Kiroku Iinkai (Ed.)
1951 Nōchi Kaikaku Tenmatsu Gaiyō (Summary of the
Circumstances of the Land Reform). Tokyo: Ministry of
Agriculture.

Nomura, Iwao
1937 Kankō Kosakuken Ni Kan Suru Kenkyū (March Site
Survey). Cited in Miyagi-ken Hasama Tochi Kairyoku
Jimusho, ed. 1965.
Nōrinsuisanshō (Ministry of Agriculture, Forestry, and Fisheries)
1982 Tensaku no Genjō (The Crop Diversion Situation). Tokyo:
Ministry of Agriculture.
1983 Nōgyō Roppō (The Six Agricultural Laws). Tokyo: Gakuyō
Shobō.
Nōrinsuisanshō Daijin Kambō Kikakushitsu (Planning Office of the
Minister's Secretariat, Ministry of Agriculture, Forestries
and Fisheries)
1986 Nijūisseki e Mukete no Nōsei no Kihon Hōkō (The Basic
Direction of Agricultural Policy Aimed at the Twenty-first
Century. Tokyo: Sōzō Shobō.
1987 Zusetsu Nijūichi Seki e no Nōsei no Tenkai (The
Development of Agricultural Policy for the Twenty-first
Century Illustrated). Tokyo: Chikyusha.
Nōrinsuisanshō Keizaikyoku Tōkei Hōkokubu
1955-85 Nōgyō Sensasu (1985 Agricultural Census). Todōfuken-
betsu Tōkeisho No 4. Miyagi-ken: Nōrin Tōkei Kyōkai.
(The Miyagi-ken individual household reports are stored at
Tōhoku University).
Nōrinsuisanshō Kōzōkaizenkyoku
1982 Hojōseibi Jigyō Benkan (Manual for the Project to
Reorganize the Paddy Fields). Tokyo.
Nōrintōkei Kyōkai
1986 Nōgyō Hakusho (White Paper on Agriculture). Annual.
Tokyo: Nōrintōkei Kyōkai.
1986 Nōgyō Hakusho Fuzoku Tōkeikyō (Statistical Appendix to
the White Paper on Agriculture). Nōrintōkei Kyōkai.
Nōsei Shingikai (Agricultural Policy Deliberation Council)
1980 Hachijū Nendai no Nōsei no Kihon Hōkō (The Basic
Direction of Agricultural Policy in the 1980s). Nōsanbutsu
no Juyō to Seisan no Chōki Mitōshi (The Long-term
Outlook on the Demand and Production of Agricultural
Products). Tokyo: Nihon Nōgyō Shimbun.
1982 Hachijū Nendai no Nōsei no Kihon Hōkō no Suishin ni
Tsuite (Regarding the Promotion of "The Basic Direction
of Agricultural Policy in the 1980s"). Tokyo.
Nōsonchiiki Kōgyō Dōnyu Sokushin Sentā (Center for the
Promotion of Rural Area Industrial Development)
1987 Endakanado ga Chiiki to sono Kōyō ni Oyobosu Eikyō ni

Kansuru Chosa Hokokusho (Survey Report Regarding the Effect of Yen Appreciation on Regions and Local Employment). A Survey Prepared for the Cabinet Policy Deliberaton Office of the Cabinet Secretariat 1986.

Oizumi, Kazunuki and W. Sugawa
1981 Miyagi-ken Nangochogai. Tochi to Nogyo (Land and Agriculture) 12:115-124.

Ono, Takeo
1977 (original 1926) Eikosakuran (The Theory of Permanent Tenancy). Tokyo: Nosangyoson Bunka Kyokai.

Ouchi, Tsutomu
1989 Nogyohogo no Tetsugaku (The Philosophy of Agricultural Protectionism). Ekonomisuto January 17:62-88.

Sankei Shinbun Henshukyoku ed.
1987 Nogyo Kakumei (Agricultural Revolution). Tokyo: Fusosha.

Sato, Tosaburo
1987 Mura ni Kita Firipin no Hanayometachi (The Philippine Brides Who Came to the Village). Ashita 6:39-44.

Shinnosei Kenkyukai
1981 Hachijunendai No Nosei No Kihonhoko. (The Basic Direction of Agricultural Policy for the 1980s). Tokyo: Osei Publishing Co.

Sorifu Tokeikyoku
1960-1985 Kokusei Chosa Hokoku (Population Census of Japan). Tokyo.

Tadano, Naosuke
1970 Tadano Nojo no Genjo (The Situation of the Tadano Farm). Miyagi-ken Tajiri-cho.

Tachibana, Takashi
1980 Nokyo (The Agricultural Cooperative). Tokyo: Asahi Shimbunsha.

Tamayama, Isao
1948 Sendai Han ni okeru Tochi Seido (The Land System in the Sendai Domain). In Tochi Seidoshi Kenkyu (Research on the History of the Land System), Yoshiji Nakamura, ed. Tokyo: Hoe Shobo.

Tobata, Seiichi, ed.
1978 Nihon Nogyo Hattatsushi (The History of the Development of Japanese Agriculture). Tokyo: Chuo Koronsha.

Tohoku Noseikyoku, ed.
1980 Yamagata Norin Suisan Tokei Nenpo (The Statistical Yearbook of Agriculture for Yamagata Prefecture). Yamagata: Yamagata Norin Tokei Kyokai.

Tomegunshi Henshūinkai
1924 Tomegunshi (The History of Tome County). Sendai: Tōhoku Insatsu Kabushikikaisha.
Toyohara Kenkyūkai
1978 Toyohara Mura: Hito to Tochi no Rekishi. (Toyohara Village: The History of Man and Land). Tokyo: Nōrinsuisanshō Nōgyō Sōgo Kenkyūsho and University of Tokyo Press.
Tōyō Keizai, ed.
1983-1987 Shūkan Tōyō Keizai. Tokyo.
Usami Shigeru
1981 Ine no Tansaku Chitai (The Rice Monocropping Region). In Tōhoku Nōgyō. Shusaku Nishida and Kanichi Yoshida, eds. pp.12-31. Tokyo: Nōsangyoson Bunka Kyōkai.
Yanagita, Kunio, ed.
1951 Minzokugaku Jiten (Dictionary of Folklore). Tokyo: Tokyodō.
Yomiuri Shimbun
1989 Nagano Sanku Dake Ichigen (Nagano District 3 Is the Only District to be Eliminated). Mainichi Shimbun, February 3.
Yoshida, Kanichi, ed.
1975 Kōdokeizaiseichō to Chiiki no Nōgyō Kōzō (The Structure of Rapid Economic Growth and Regional Agriculture). Tokyo: Nōsangyoson Bunka Kyōkai.
1981 Kazoku Keiei no Seisanryoku (The Productivity of Family Management). Tokyo: Nōbunkyō.
Yuize, Yasuhiko
1987 Kore dewa Nogyo Kaikaku wa Susumanai (Reform in Agriculture is Getting Nowhere this Way). Ekonomisuto August 31:96-108.
Zenkoku Nōgyō Kaigisho (National Land Commission Office)
1980 Suiden Kosakuryō no Jittai ni Kan Suru Chōsa (The Survey on the Situation of Paddy Tenancy Rent Rates). Tokyo: Zenkoku Nogyōkaigisho.
1982(a) (Revised 1986) Me de Miru Nogyōsha Nenkin (Agricultural Retirement Fund Picture Book). Tokyo: Zenkoku Nōgyōsha Nenkin Renraku Kyōgikai.
1982(b) Wakariyasui Nōchi no Zeisei (The Agricultural Land Tax System Made Easy). Tokyo: Zenkoku Nōgyōkaigisho.
Zenkoku Nōgyō Kyōdō Kumiai Chuōkai (Central Committee of the National Agricultural Cooperative Associations)
1987 (a) Nōkyō Nenkan (The Yearbook of Agricultural Cooperatives). Tokyo: Zenkoku Nōgyō Kyōdō Kumiai

Chuōkai.
1987 (b) <u>Gatto to Nōgyō</u> (GATT and Agriculture). Tokyo: Zenkoku Nōgyō Kyōdō Kumiai.

WORKS IN CHINESE

Peng Tsu-Kwei
1987 Taiwantichu Nungchang Kuimo P'ientong chi pouxi (The Changes in Farm Size in Taiwan). <u>Nungyeh Chinchi Panyuhk'an</u> (Agricultural Economics: Semiannual Publication of the Research Institute of Agricultural Economics of National Chung Hsing University) 41:1-14.

Index

334